The Internet and Business English

Barney Barrett

Pete Sharma

Summertown
Publishing

Published by
Summertown Publishing Limited
29 Grove Street
Summertown
Oxford OX2 7JT
United Kingdom

E-mail: info@summertown.co.uk
Web-site: http://www.summertown.co.uk

ISBN 1902741773

© Summertown Publishing Limited 2003

First Published 2003
Reprinted 2005

Editor: Susan Lowe

Authors: Barney Barrett and Pete Sharma

Page design and setting: Oxford Designers & Illustrators

Cover design: Peter Mays, Whitespace

Printed in Spain

Contents

About the authors

Pete Sharma

Pete is the Group Teacher Training and Development Manager for Linguarama International, a language training organisation running courses for Business English learners primarily in Europe. Pete has been with the company since 1980, has taught mainly in Spain and Finland, and is currently based in Stratford-upon-Avon, UK. He runs teacher training courses in-house and has been a trainer on the LCCI Dip TEB and UCLES Dip BPE.

Since 1994, Pete has reviewed CD-ROM discs and given a great many presentations on CD-ROM at various conferences, notably BESIG (Business English Special Interest Group) and IATEFL (International Association of Teachers of English as a Foreign Language), as well as TESOL (Teachers of English to Speakers of Other Languages) and JALT (Japanese Association of Language Teaching).

Pete has been a BESIG Committee member and has edited a number of BESIG newsletters, for which he regularly contributes the CD-ROM / Internet page. He also contributes articles and reviews to EFL magazines such as *IATEFL Issues*.

He is the author of *CD-ROM: A Teacher's Handbook* (1998, Summertown Publishing) updated as an e-book (2002).

Pete is currently studying on a distance-learning M.Ed course in Technology and ELT at Manchester University, and has taken the COLTE (Certificate of On-Line Teaching of English) course at NetLearn Languages. He is particularly interested in e-language learning and ways in which technology can support the learning of languages.

Barney Barrett

Barney also works for Linguarama International. He started with the company in 1994 and has taught in Spain and the UK. Barney has extensive experience in teaching Business English to individual learners and groups.

Since 1996 he has been based at Linguarama's Stratford-upon-Avon centre where he has gained a wealth of experience in using computers in language teaching, greatly assisting the teaching staff in integrating technology into Business English courses.

Barney has long been interested in technology and the Internet, as well as their application in language teaching. He helped devise and pilot material for *CD-ROM: A Teacher's Handbook*, and has worked on in-house teacher training sessions on using CD-ROM, the Internet and e-mail. He has also been instrumental in the design and programming of two in-house teacher training multimedia CD-ROMs.

Foreword

The Internet plays a key role in business today. Most businesses, even small ones, have a web-site; e-mail has overtaken fax as the preferred method of written correspondence.

Any teacher of Business English needs to be Internet-aware, not just because of the importance of the technology in the work of most of his/her students, but because the Internet is changing the ways in which we can teach English, and indeed is changing the language itself.

Computers have been used in language learning since the Sixties, but it is only in the last decade or so that their use has moved into the mainstream. The Internet can be used for all the 'conventional' computer-assisted activities of the past, but with two important additions: access to an inexhaustible range of language material, and the ability to communicate with other users around the world.

With such rapidly changing technology, so many web-sites, and so many potential pitfalls, the Internet can be daunting for any teacher to get to grips with. Surfing the Web is easy; using it effectively and efficiently to create motivating activities is more of a challenge.

This book is a practical introduction to the Internet for teachers of Business English, and a useful guide to one of the most exciting areas of development in English language teaching today.

David Eastment

Introduction

The technical revolution has changed the nature of business practice in innumerable ways: the rise of e-commerce, the use of e-mail as a form of communication and the growth of company web-sites to name but three.

It has inevitably affected the teaching of Business English in a number of ways. Firstly, it has influenced what we teach: vocabulary items such as neologisms; how to compose e-mails; learner training in the use of the Internet and so on. Secondly, it has influenced how we teach: the rise of web-based training materials and CMC (Computer-Mediated Communication) are additional possible teaching tools.

The Summertown series

This book is part of a series of Handbooks written to support language teachers, focusing on specific areas. The first book in the series was *CD-ROM: A Teacher's Handbook* (Pete Sharma, 1998, Summertown Publishing).

The Internet and Business English looks in particular at the use of the World Wide Web in Business English teaching. It considers ways of using the Web in face-to-face teaching or for self-study. It does not focus on on-line teaching.

Who is this book for?

This book is primarily for Business English teachers. However, it will also be of use to those in pedagogical management such as Directors of Studies, and Heads of self-access centres.

Users of the book may be experienced in using technology such as the Internet, or be completely inexperienced. They may be highly experienced Business English teachers, or relatively new to this field, perhaps having recently diversified from General English teaching. They may have a background in the business world or they may not. They may work for a Business English school or on a freelance basis. They may be based in a school or in-company.

What is Business English?

There are a number of defining characteristics of Business English. These include:
- a needs analysis, usually undertaken before a course
- a course programme based largely on the language and skills needs of the learner or group of learners
- a focus on common business skills areas such as making presentations and participating in meetings
- lesson content based on using the learner as a resource

- the use of authentic teaching materials provided by the learners, such as company brochures
- a Learning to Learn element
- learners on one-to-one courses, not only in groups.

An overview of this book

The book is divided into three parts: The Internet, Practical ideas and Reference. It is also supported by information on the Summertown web-site.

Part one - The Internet

Chapter 1 looks at the Internet and covers essential technical features and vocabulary. It is essential reading for the Business English teacher relatively new to technology. It can be skipped by those with a good technical knowledge. However we advise skimming these pages as they also outline reasons for the various technical choices the authors have made concerning browsers etc.

Chapter 2 links the Internet to language teaching, examines why we should use the Web in teaching and how to do so.

Chapter 3 looks at how various aspects of Business English teaching can be practised through integrating the Web. It examines the four language skills of speaking, listening, reading and writing; a range of common business skills such as telephoning and giving presentations, and other areas of Business English such as inter-cultural training and examinations.

Chapter 4 explores how the Business English teacher can gain professional support and development through using the Internet, through forums and readily available materials.

Part two - Practical ideas

Many learners need help and guidance with Web skills and Chapter 5 looks at learner training. It offers practical ideas for helping learners become familiar with terminology and with using the Internet.

Chapter 6 offers a range of practical teaching ideas. These are multi-level tasks, and each task either incorporates the use of the Web, or has a Web-based theme. There is a bank of grammar practice activities, tasks designed for vocabulary enrichment, practising business skills and fluency. Many practical ideas are supported by framework material and questionnaires. These are included in the book, and are also available on the Internet (see over). It is envisaged by the authors that teachers will adapt the tasks and material to their own teaching situations.

Chapter 7 is for Directors of Studies and Teacher Trainers. It provides a bank of teacher training ideas to help develop Business English teachers in the area of the Internet, and enable them to integrate the Web into their teaching programmes.

Part three - Reference

Chapter 8 is a data bank divided in two sections. Section 1 is a data bank for specialist areas, such as insurance, finance and law. Within each subject area, the authors suggest key sites which are recommended for a number of reasons: as a source of articles which may be of interest to learners in that particular field; as a source of information to be used by these learners after they have finished their language course; and as a stimulus for practical tasks. Section 2 comprises a useful list of Web addresses for major companies. This is designed to help teachers access sites which may prove helpful in their teaching.

Chapter 9 offers a range of useful tips for teachers wishing to incorporate the Internet into their teaching, such as how to download a graphic or plug-ins. It also offers technical advice for common problems.

Finally, Chapter 10 provides two glossaries. The first is a list of technical terms such as *ISDN* or *cache*. The second is a list of pedagogical terms such as *blended learning*. The first occurance in the text of glossary items is given in blue.

On-line support

The book is currently available in a printed format, with support provided on the Internet. Web-sites are updated or changed with alarming frequency, and information on any changes to addresses will be given on the Summertown web-site. See: http://www.summertown.co.uk

Users will be able to access the following:
- PDF versions of the worksheets for printing out
- extended reference sections with hyperlinks, allowing immediate access to the web-sites
- bulletins on technology, which will keep the book up-to-date
- further practical ideas submitted by Business English users across the world.

And finally...

There are a number of excellent books for language teachers about the Internet and these are listed in the bibliography. However, *The Internet and Business English* specifically addresses the area of Business English, making it different in both content and approach. We hope you enjoy using this book.

January 2003

Part one
The Internet

1 What is the Internet?

Chapter 1 is divided into the following areas:

1 The Internet
2 The World Wide Web
3 Intranets and extranets
4 Getting connected to the Internet
5 Web-sites
6 Browsers
7 Searching the Web
8 Web audio and video
9 The Internet and CD-ROM

1 The Internet

Before going into detail of how the Internet can be used in Business English teaching, a useful starting point is to clarify what the Internet is. At the end of the book is a glossary, in the meantime, there are a few terms that will be used on the intervening pages which need to be explained from the outset.

The Internet is a global network of interconnected computers which communicate with each other using a common computer language (TCP/IP). Originally set up as a decentralised military communications system designed to survive a nuclear attack, the network was later opened up to academic, then public and finally, commercial use.

The Internet is therefore the infrastructure upon which the World Wide Web, e-mail and other forms of Computer Mediated Communication (CMC) rely.

If you wish to explore these topics further, you may find the following web-sites useful.

Definitions:

http://whatis.techtarget.com/definition/0,289893,sid9_gci212370,00.html

History:

http://www.zakon.org/robert/Internet/timeline/

http://www.w3.org/History.html

http://www.netvalley.com/intval.html

2   The World Wide Web

The World Wide Web (the Web or WWW) has been in existence since 1994. It was invented by a man called Tim Berners-Lee and is the part of the Internet where websites and their component web-pages are stored and can be accessed. Tim Berners-Lee described the Web as: '[. . .] the universe of network-accessible information, an embodiment of human knowledge,' (taken from http://www.w3.org/WWW) which gives you an idea of the utopian ideals that the early pioneers of public Internet communication subscribed to. Today, however, the Web seems to be dominated by big business and commercial interests. These are the areas that this book is concerned with exploiting for teaching purposes.

2.1   Using the World Wide Web

When you type a Uniform Resource Locator (URL) into your browser (see Section 6 below for more detailed information) the software requests information from the computer hosting the web-site and web-page you have requested. The web-page is then delivered to your computer in packets of information. This information is coded using a computer language called HTML. Your browser reads the HTML and resolves it into the words, pictures and hyperlinks that make up the web-page you requested.

Hyperlinks are a defining element of the Web and how it works. Move your mouse pointer around a web-page. As it passes over certain words, images or graphic buttons, the pointer becomes a small pointing hand. This identifies the word or button as a hyperlink. Click on the hyperlink and you are instructing your browser to access and display another web-page within the web-site you are already in or a web-page within another completely different web-site. Hopping from web-pages and web-sites, following links that attract your interest or tweak your curiosity, is known as surfing. The physical geographical organisation of the Web is unimportant and its raison d'être. Web-pages can be stored on computers anywhere in the world. Hyperlinks then enable any two things accessible on the Web to be connected

The Web is huge and growing everyday. In January 1997 there were something in the region of 700,000 web-sites. By the middle of 2001, this figure had increased to 31.3 million! Although the Web contains hundreds of millions of items of information in the form of text, pictures, audio and video clips, there is no organisation to any of this information beyond the design of individual web-sites. There is no 'gatekeeper' controlling what can and cannot be published on the Web. Attempts to apply existing laws to offending information on the Web has resulted in various court cases around the world with varying degrees of success on the part of the prosecuting authorities. Nevertheless, you can avoid web-sites of unsavoury or unsuitable content and the benefits of the Web far outweigh any dangers. On a more practical note, finding information on the Web is not always straightforward and a skill you will need to develop. In Section 7 of this chapter, we will discuss some effective ways of searching the Web.

3 Intranets and extranets

If you work for a large organisation or teach in-company, you may have met the terms *intranet* and *extranet*. These terms require some explanation, since both nets are usually accessed through a Web browser and will probably look very much like the Web which is on the public Internet.

Both intranets and extranets are private networks that use the same type of language as the Web to code and communicate information. An intranet is limited to the internal network of a particular company, part of a company or institution. It is not accessible to outside users via the Internet. If a company has a Wide Area Network (WAN) linking several locations then the intranet will be accessible across that entire WAN.

An extranet may appear to be the same as an intranet. The main difference is that an extranet exists as part of the Web. It is protected from general access by passwords and other security software such as firewalls. Through these security protocols an extranet can be accessed via any computer connected to the Internet. This allows remote access to company information by employees and, often also, customers. There are a variety of other uses of extranets such as remote e-mail access.

4 Getting connected to the Internet

We will assume that you have access to a computer that has an Internet connection. Information on how to get on-line can be obtained from the numerous Internet Service Providers (ISP). In this section, however, we will briefly outline the different types of connections available in order to clarify issues such as download speed, cost, expectations (of yourself and your learners) and time spent on-line, i.e. connected to the Internet.

For a list of available ISPs look at: http://thelist.internet.com/countrycode.html

There are two ways of connecting to the Internet:
• via a modem
• via the office or company network that your computer is part of.

4.1 Connecting through a modem

The modem option is the one that most home-users are familiar with. In a business environment modems are now extremely uncommon. If you use the Internet at home then you are probably connecting via a modem. This is the component or peripheral that makes the screeching noise when you click on the 'Connect' button of the 'Dial-up Connection' box of your browser. The modem connects you to a server belonging to your ISP. This server is the first of many different computers that information will travel through to and from your computer.

The speed at which information is sent to and from your computer depends on a number of factors. The main ones are the speed of your modem and the quality of your telephone line. Most home PCs now come with a 56k modem installed. The 56k refers to the number of bytes per second (bps) that the modem is capable of sending and receiving; in this case 56,000. Older modems had speeds of 33,300; 14,400 and even 9,600bps. If you have one of these you need to think about up-grading. A final factor is how up-to-date your computer is. There is a simple formula: the newer the computer, the faster it is able to process information.

Speed is important because the slower your connection, the longer you spend waiting for web-pages to download and this may have a cost implication. Additionally, there are more and more web-sites and on-line facilities whose accessibility is highly limited if you have a slow Internet connection.

If you are using a computer with Microsoft *Windows* installed, a small flashing symbol will appear in the bottom right-hand corner of your screen, next to the clock while you are connected to the Internet. Hold your mouse over this symbol and your current connection speed will be displayed. Double-click this symbol and you will get a box giving the connection speed; how long you have been connected and how much information has been sent and received by your computer. These last two figures should be continually changing as your computer requests information and it is downloaded to be displayed by your browser.

Although you may have a 56k modem, technical factors mean that it never runs at 56,000bps. The degree to which it is below this figure is often determined by the quality of the telephone line you are using. The same computer system could achieve a connection speed of 52,000bps using one telephone line but only 36,000bps on another in a different building or town. This is out of your control.

4.2 Connecting through a network

If you are using a computer that is part of a network, your connection to the Internet is arranged in a different way to that of the modem user. Once the computer has been switched on, it is necessary to enter a password to identify yourself to the network. Without this you are unable to access common data or the Internet. When you open your browser you are instantly connected to the Internet and your homepage starts to download.

Because a network can be serving anything between two and several hundred computers, the system for connecting to the Internet is different to a modem. It usually involves some type of dedicated, high-speed telephone line. These lines and connections are available to home-users but at a cost which, at the time of writing, makes it uneconomical for the occasional surfer. You will hear terms such as *ISDN, ADSL* and *VDSL*. Each represents an increase in speed of access.

The term *broadband* has been employed for several years now. Since the technology is changing so quickly, last year's broadband now seems slow. Broadband therefore is shorthand for high-speed Internet access. Some potential applications are high-quality video-telephony and full-screen television delivered via the Internet

5   Web-sites

A web-site is identified by its domain name. The domain name is the part of the alphanumeric code that we use to identify the address of a web-site. This whole code is called the URL or Uniform Resource Locator. All URLs for web-sites on the World Wide Web begin with http:// followed by the domain name. For example, the URL of the US Cable News Network (CNN) is http://www.cnn.com/. The domain name is www.cnn.com. All the web-pages that make up that web-site begin with this address. Therefore, the address of the CNN web-site's audio page is http://www.cnn.com/audio/ and so on.

One common analogy for a web-site is a book. It is made up of pages with the front page usually featuring some sort of contents list. The analogy stops here; unlike a book, the pages of a web-site are very rarely intended to be read in a linear manner. On the contrary, very often every page of a web-site is connected or linked to every other within that web-site via a system of hyperlinks, menus and search facilities. A web-page may also contain links to web-pages which form part of another, completely different web-site.

Another way of thinking about a web-site is as a collection of interlinked web-pages whose coherence comes from a consistency of the topics covered by the contents and / or the design of the web-pages. Most web-sites are built around a particular subject or service. Very often, the domain name of that web-site is intended to reflect that subject or service. Therefore, CNN's web-site is called cnn.com and the BBC's is called bbc.co.uk etc.

The design and usability of web-sites varies dramatically. This ranges from web-sites where the way to find the information you want is immediately obvious to web-sites where exploration and discovery are the aim of the designers and little is done to help or guide you. There are no official rules for the design and layout of web-pages. As you become more experienced in navigating your way around web-sites, you will be able to spot which ones are well designed and which ones are not. This will also become a factor in your choice of web-sites to use with learners, or even, the basis of a classroom activity. In Chapter 9 there are some suggestions of how to read common types of web-site in order to be able to navigate to the information you want as quickly and smoothly as the web-site allows.

6   Browsers

The Web is accessed via a piece of computer software called a browser. If you use a *Windows* PC, you will have a version of Microsoft's *Internet Explorer* (*IE*) already installed and ready to use. If you are an Apple *Mac* user, see Chapter 9 for information about what Microsoft offer for your computer. The next most commonly used browser is *Netscape*. Both Microsoft *Internet Explorer* and *Netscape* are free and their regular up-grades easy to acquire. In addition, there are other products on the market, such as *Opera,* which cost a nominal amount and there are proprietary browsers such as the one provided by AOL to its subscribers. However, since we envisage the majority of users of this book will be accessing the Web via computers within an institutional or office environment, it seems very unlikely that this will involve using something like the AOL browser which is a piece of software intended for home-use.

Figure 1.1
Microsoft Internet Explorer 6

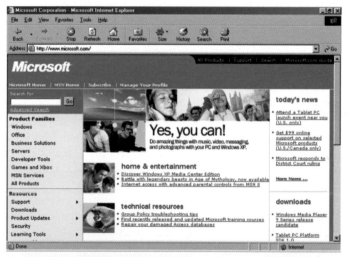

Figure 1.2
Netscape 7

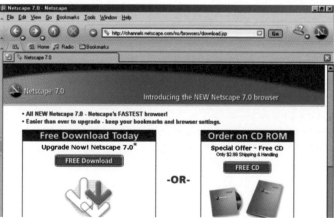

6.1 Browser version

All software companies regularly up-date their software in order to introduce new features; fix old problems and accommodate new file formats and standards. Every version of a piece of software is identified by a number, e.g. *Internet Explorer 5.5*, *Netscape 6* etc.

This book is going to assume that you have and use version 4 or later of Microsoft's *IE*. The reasons for this are:

- *IE*'s virtual omnipresence means that a huge number of existing Web users, including many Business English learners, are familiar with its interface and functions. It is also more likely to be available for use than other browsers.
- the Web is a world in which the battles over standards continue. Different browsers from different companies interpret the HTML code they are sent differently. This means that a web-page could appear correctly when viewed through one browser but completely scrambled when viewed through another. *IE* being the most common browser at the moment means that many web-site designers use it as a standard in order to accommodate the majority. The minority using other browsers may occasionally miss out.
- the majority of web-site designers subscribe to a concept called 'backwards compatibility'. This means that their sites can be accessed and used by Web users with older browsers. There is, however, a limit to how far back most designers are prepared to go. So, once again, for the purposes of this book, we are going to assume that you have *IE 4* or higher. [1]

[1] Like its competitors, Microsoft up-dates *IE* on a regular basis. At the time of writing the current *Windows* version is *IE 6*, although *IE 6.5* is not far off. The version you are able to run on your computer is limited by the version of *Windows* you have installed. If you have an old computer that runs *Windows 3.1* then you are limited to *IE 3*. If you have *Windows 95* as your operating system then you can progress as far as *IE 5.5*. If you have *Windows 98, 2000* or *XP* then you can run the latest versions of the software. See Chapter 9 for details about Apple *Mac* versions.

6.2 Using a browser

At the top of the *Internet Explorer* (*IE*) window are several menus similar to those you would find in most other programs; a series of buttons and then a text entry box labelled 'Address'. There might also be another line of buttons below that.

The following is a tour through the main functions of *IE*. You may also want to refer to the glossary at the back of book.

Address | In order to discuss the functions of the browser, it is useful to look quickly at how it works. Once the browser is open and an Internet connection is established, you enter a URL into the box labelled 'Address' and then press 'Return' on your keyboard or click on the 'Go' button on the right of the 'Address' box.

Figure 1.3
'Address' box

Your browser now sends out a signal that it requires the information which makes up the web-page whose address you have entered. If you have entered the address of an existing page, the browser will establish a connection with the computer hosting that web-page and download the information to your computer via the phone line. The browser interprets that information and converts it into the text, pictures and graphics you see in the browser window. Remember that URLs are case sensitive. If you type in a URL that does not exist or you make a mistake as you enter the URL, the browser will be unable to make the connection and all you will receive is an error message in the window.

Figure 1.4
'The page cannot be displayed' message

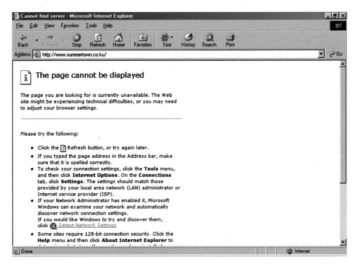

Back and forward | Once the web-page has been downloaded and is displayed in the browser window, you can use your mouse to click on a hyperlink to take you to another part of that web-site or to another page on another web-site. When you do this, the new web-page you have requested replaces the first one in the window and the same things happen if you then click on another hyperlink and then another. However, your browser remembers where you have been. At any point you can click on the 'Back' button at the top of the browser window to return to the previous page. Clicking on the 'Back' button again will take you back one more page and so on. Conversely, clicking on the 'Forward' button will reload the page you visited after the one you have just returned to.

By using the 'Back' and 'Forward' buttons you are accessing the 'History' of the Internet session you are currently engaged in. This function means that you do not have to remember the address of every page you have been to during that on-line session.

Figure 1.5
'Back' and 'Forward' buttons

History | As you look along the line of buttons at the top of the browser window you will see one that is labelled 'History'. This function allows you to access a record of all the web-pages you have visited that day, the day before or even in the previous week.

However, viewing the 'History' this way gives you a list of the web-pages sorted according to their domains rather than the order in which you visited them. Once you have finished an on-line session and closed the browser, the 'History' stored by the 'Back' and 'Forward' buttons is lost, only the record described above being retained. A further use of the 'History' function is browsing off-line. (For details see Chapter 9.)

Figure 1.6
'History' list

Stop | The next button is the 'Stop' button. This interrupts the process of downloading a web-page. The larger the web-page (in terms of graphics and other information) the longer it takes to download into your computer. It may be the case that you realise early on in this process that this web-page is of no interest to you. You can, therefore, click on the 'Stop' button to halt the download and use the 'Back' button to return to the previous page in order to try another hyperlink. As Internet connection speeds get faster and faster web-pages sometimes download so quickly that there is no time to interrupt them. Consequently, the 'Stop' button is set to become redundant in the near future.

Figure 1.7
'Stop' button

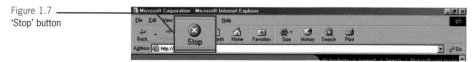

Refresh | Next to the 'Stop' button is the 'Refresh' button. When this button is clicked the browser is instructed to download the most up-to-date version of the web-page currently being displayed or whose URL is entered in the 'Address' box. Usually this involves renewing the Internet connection and downloading the web-page. Many web-pages do not change from one month to the next; others, such as news web-pages, can be up-dated every hour of the day and night. The way browsers are currently programmed means they will not seek out up-to-date versions of a web-page unless you tell them to. For example, if you downloaded the homepage of the BBC News web-site, you would see the very latest headlines. If, after twenty or thirty minutes, you clicked on the 'Refresh' button, your browser would reload the page and the headlines may have changed or a new top story may have been introduced. However, for the time being, your browser has no way of checking whether changes have taken place on a web-page and providing you with an up-dated version of the page automatically.

Figure 1.8
'Refresh' button

Home | The next button is labelled 'Home'. *Homepage* is a potentially confusing term. Every web-site or web domain has a homepage. This is sometimes also called the frontpage of the web-site. This is where you usually find the main menus or an introduction to the web-site and its contents. On a browser, the homepage is something a little different. When you open your browser it automatically goes to a pre-programmed web-page. This is your homepage. It is very simple to set this default page for your computer's browser. Most people set their homepage to a web-page they use very often. For example, if you use the search engine *Google* virtually every time you connect to the Web, then it would be useful to have it set as your homepage. To do this, click on the 'Tools' menu at the top of the browser window and then click on 'Internet Options' at the bottom of that menu. A dialogue box opens and the very first item is the homepage. In the text entry box type *Google*'s URL: http://www.google.com then click on 'OK' at the bottom of the dialogue box.

Figure 1.9
'Internet Options'
dialogue box

Now, whenever you open your browser or click on the 'Home' button, you will be taken straight to the *Google* homepage. You can change your browser's homepage as and when you change your Web habits.

Favourites | 'Favourites' are web-pages that have been bookmarked (the word that Netscape uses) using your browser in order to facilitate quick access to them next time you wish to visit those web-pages. All browsers allow this type of activity. A useful feature of 'Favourites' is that you are able to mark individual web-pages. You are not limited to the homepage of a web-site.

The original idea of 'Favourites' was to mark the web-sites or web-page that you visited regularly to save typing in their URLs each time. With the rapid growth of the Web, the facility is equally useful for storing URLs of web-sites and web-pages that are interesting either for use now or in the future. Often these web-pages have been found accidentally through surfing or sorting through the results of a search. Marking these web-pages as 'Favourites' there and then ensures that you will be able to find them again if or when they meet a need. Each individual 'Favourite' file is tiny in terms of memory usage. Therefore, there is no need to worry about marking everything you encounter that seems interesting as a 'Favourite'. However, as this list builds up organisation becomes vital.

Like any system for filing, storing and retrieving information, you need to start as soon as possible and then maintain it. Organising 'Favourites' is easy. A new 'Favourite' is created by clicking on the 'Favorites' menu in the browser window and clicking 'Add to Favorites'. This adds the web-page that you have open to your 'Favourites' list. However, it is possible to build a system of folders in which to store your 'Favourites'. Like the system of folders that you use to store word processor files, graphics or any other type of digital information, the way you chose to organise it is up to you and should match your priorities and work practices.

Making folders for 'Favourites' is very straightforward. Select 'Add to Favorites' and the following dialogue box will open.

Figure 1.10
'Add to Favourites'
dialogue box

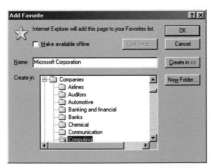

Click on 'Create in <<' to reveal the list of folders then click on 'New Folder'. Give the new folder a name, make sure it is highlighted and click on 'OK'. The 'Favourite' will be created in that folder. You can then create other folders or folders within folders.

An example of this can be seen in the language centre where we work. The computers have a folder in their 'Favourites' list called 'Companies'. Inside this folder are a number of other folders called 'Auditors', 'Automotive', 'Banking', 'Chemicals', 'Consulting' etc. Inside each of these folders are the 'Favourites' linking to the homepage of our major clients in each of these business sectors. Other folders in the 'Favourites' list include 'On-line dictionaries', 'E-mail writing', 'Publishers and bookshops'. The time saved by both teachers and learners who use this list is significant.

The main folder containing the 'Favourites' list called 'Favourites' is stored inside the 'Windows' folder on your computer's hard drive. The contents of this can be copied to other computers either across a network or using a floppy disc. Be careful not to delete your list while doing this. Since a 'Favourites' list has often been built up over a period of years, it is a good idea to make a regular backup on a floppy disc, zip cartridge or CD-R and keep this in a safe place.

Other buttons | The operation of the 'Print' button seems logical but does not always produce the results you expect. It is often better to click on the 'File' menu and then select 'Print' from there. The dialogue box which then appears gives you more control over what part of the current web-page you are actually sending to the printer and how it will be printed. *IE* version 5.5 and higher has a 'Print Preview' feature. This allows you to see whether the web-page will print on portrait-orientated paper or whether you need to reset it to landscape. The 'Size' button can be useful. This button gives you a small menu with a choice of five text sizes. Making text larger can help when reading from the screen, especially if you only have a small monitor. It is worth noting that this button does not always have an effect. On many web-sites the text size has been fixed by the designer and cannot be altered using browser controls. The 'Search' button should not only be treated with caution but probably best ignored. It activates Microsoft's own Web search tools. There are other more efficient ways of finding what you want on the World Wide Web. (See Section 7 below.)

Figure 1.11
'Search', 'Size' and
'Print' buttons

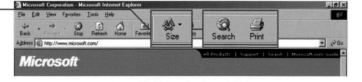

Searching the Web

Searching the Web can sometimes be a frustrating and fruitless task. There are three things to keep in mind before starting your search:
• be clear about what you want to find. Be as specific as you can. As we will see, the more specific or focused your search, the fewer the results you will have to look through and the faster you will get where you want to go.

- select an appropriate tool. Understanding how search engines and directories work, find and classify information; how they interpret the information you enter into the text search box and how they display the results will help you to choose the best tool to use.
- ask clear, detailed questions using the language your chosen tool understands.

7.1 Content

The Web is often described as being like a library or an encyclopaedia. However, this comparison is misleading. Firstly, unlike a library and an encyclopaedia there is no central control over what is put on the Web. There is no control over its accuracy, its up-to-dateness, its truthfulness or any of the qualities you would expect of information stored in a system with some sort of gatekeeper. Secondly, all this information, stored in more than 35 million web-sites, some of which consist of many thousands of pages, is not organised in any way other than that chosen by the authors of the individual web-sites. The World Wide Web has no index and no contents page. There is no central place in which you can find listings for the entire Web.

However, it is still possible to find what you want on the Web. Bear in mind that all of human knowledge is not there. What is on the Web is what people and organisations have chosen to put there, for whatever reasons. However, if it is there and general access is intended, i.e. not protected by passwords, it is possible to find it. The following section talks in some detail about searching the Web. Before embarking on this, the most important question you need to answer is: What *exactly* do I want to find on the Web?

7.2 Search engines and directories

Since finding anything on the Web is virtually impossible on your own, help is at hand in the form of search engines and Web directories. These services are funded by advertising and are free to use. However, the way they process your query and create a list of results is very different. Understanding this will help you to decide which to use in what circumstances and, therefore, make your searching very much more efficient and successful.

Search engines | Search engines examine the World Wide Web by sending out programs called spiders or crawlers to read the HTML code of web-pages located on the Web. The information that is collected is then used to create a catalogue called an index. When you use a search engine it searches its own index. Different search engines have different sized indexes based on the type of spider software they use and on the rules which determine what information those spiders bring home. At the time of writing, the largest index belongs to *Google* with over 3 billion web-pages. However, a search engine cannot inspect and catalogue 3 billion web-pages in an instant. This means, at any one time, various parts of the index are more up-to-date than others. This is why search engines sometimes throw up dead links, which means the web-page in question has expired or been deleted in the time between the spider adding it to the index and the search engine offering it up to you as a result to your search. (For more information see Chapter 9.)

Examples of search engines include: *Google, Altavista,* and *Hotbot.*

Directories | Directories examine databases of web-site URLs that have been created and are kept up-to-date by the directories' human operators. Every web-site has been selected, categorised and had some sort of short description added. Therefore, when you run a search in *Yahoo,* for example, the results come from *Yahoo*'s own, extensive directory. Some web search services combine the two, often by simply submitting your query to several other search engines or directories then summarising the results on a single output web-page.

Examples of directories include: *Yahoo!, Lycos, Ask Jeeves* and *Excite.*

When to use a search engine or a directory | Clicking on a hyperlink from a search engine's result list takes you to the exact page on which the words and phrases you are looking for appear, regardless of whether it is the homepage of the web-site or not. You should use a search engine when you want to get to a specific piece of information quickly.

Clicking on a hyperlink from a directory's result list generally takes you directly to the homepage of a web-site. From there you can find what you want using the web-site's own navigation. If you have a less specific search target, use a directory. Since the listings are organised by category, you can get where you want to go in stages.

The following table gives some general examples of types of searches and explains when it is better to use a search engine or a directory.

Figure 1.12
When to use a directory or a search engine

Use a search engine...	Use a directory...
to find out which companies manufacture a particular type of product.	to find the web-site of a particular company.
to find an exact quote from a newspaper report or in what type of stories a word is used.	to find a list of newspaper web-sites categorised by topic or country or language.
to find the context in which a piece of specialist jargon is being used.	to find specialist dictionaries or glossaries.
to find examples of common mistakes on web-pages and in discussion forums.	to find web-sites which discuss common language mistakes.
to find examples of specific types of written documents or correspondence.	to find web-sites that give advice on effective writing.

7.3 Effective searching

The following are some common searching scenarios and how we would go about tackling them to find what we need.

Finding a company's website | In this case our learner was a buyer for the Japanese retail chain Ito-Yokado and we wished to incorporate the web-site into his lessons (see Chapter 6: Lesson Activities). Often the first, and quickest strategy is to use the company's name and add '.com'. Earlier in the course he had wanted to see the web-site of the British supermarket chain Tesco. Entering http://www.tesco.com into the address box of our browser had taken us straight to the correct web-site and from there it was simple to find links to the corporate pages and those geared to customer services.

With this in mind we tried http://www.ito-yokado.com. Although this web-site address does exist it is owned by a company which sells URLs. We therefore tried a variation which is common for Japanese companies http://www.ito-yokado.co.jp. This address did not go anywhere. Our next stop was the Web directory *Yahoo!*. We entered the name of the company into the search box and results came back with the address http://www.itoyokado.jyg.co.jp/iy/index.htm.

This brought up the homepage of Ito-Yokado, but in Japanese. Clicking on a hyperlink labelled 'English' took us to http://www.itoyokado.jyg.co.jp/iy/index1_e.htm, the English language version of the web-site and exactly what we required for our lessons. Success!

Finding a list of company web-sites from a particular industry | For the activity in Chapter 6 which involves checking flight times on airline web-sites, we needed to make a list of web-sites. Rather than brainstorming names of airline companies then finding their web-sites one by one, we went straight to *Yahoo!* and entered 'airlines' into the search box. From the 'Category Matches' we clicked on 'Airlines' and *Yahoo!* gave us an alphabetic list of all the airline companies in the world which have web-sites. From there it was straightforward to make a selection of the ones we wanted to use for the activity.

Figure 1.13
List of airline
company web-sites
from *Yahoo!*

Finding a specific topic related to a learner's business / job | The next search was for a very specific topic. The learner worked for the food company Nestlé and wanted some information about the company's yoghurt in Spain. A journey through the company's extensive web-site revealed that the information we wanted was either not there or not accessible to non-employees. This time we turned to a search engine. In this case we used *Google*.

A basic rule with search engines is to give them as much information as possible. The more words you enter into the search box, the more closely focussed your search will be and the more likely it is that you will find what you are looking for. For this search we entered 'factory production yoghurt spain nestle'. *Google* provided us with a list of seventeen results.

Figure 1.14
Search results from
Google

A list this short made it very straightforward to visit every web-page to check it for the information we required. Many of the pages contained food industry information and consisted of large amounts of text. By pressing 'Ctrl+F' on the keyboard we opened the 'Find' box and were able to search through the text quickly. This allowed us to rapidly discover whether references to Nestlé in the text were linked to yoghurt production in Spain or not. The following web-page provided the information we required: http://www.perlarom.com/gb/news/spanyoug.html

Finding an explanation of a grammar topic | The next search is for an on-line explanation of an aspect of English grammar. In this case we wanted something on modal verbs. It shows how we made use of search engine maths, i.e. adding and subtracting terms in the search box to narrow our search and reduce the number of results to something more manageable.

First we entered 'english grammar modal verb', which produced about 8,330 results. We were interested in modal verbs for obligation so we added 'must' to our search and the number of results fell to 5,930. Adding 'should' brought it down a little further to 5,320. This was not surprising since most web-pages dealing with *must* in a grammar context also include *should*.

Skimming through the results from the previous search revealed a lot of references to German grammar and how modal verbs translated between English and German. In order to eliminate these results we added '-german' to the search. The minus sign is understood by *Google* to mean exclude any page that fits the criteria of the search but which includes the attached word. The number of results fell to 2,920.

Previous experience has shown that a significant proportion of web-sites offering help and explanations of English grammar are aimed at native-speakers who wish to improve the quality of their writing. The vast majority of these web-sites have been produced by and for Americans. In order to try and filter out those pages the simplest way was to add '-written -writing -write' to the end of the search. The number of results fell to 635.

Figure 1.15
Google search results

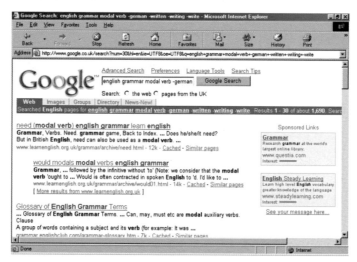

Although 635 is still a large number, we stopped there since we found enough examples for our purposes in the first 60 results by scanning the results list and only opening those web-pages that looked useful. However, we could have continued to exclude web-pages from our search by looking for regularly occurring words that did not fit the criteria we had in mind. For example, our list still contained web-pages giving advice on translation. Excluding words like *translate, translation, translator* etc. could have further reduced the number of results.

Finding a specific phrase | In the previous two searches we added or subtracted words. The search engine then looked for web-pages that included those words but not necessarily together or adjacent to each other as we saw with the yoghurt

2 The initial stimulus for using this type of searching in language teaching came from *The Internet* (S. Windeatt, D. Hardisty, D. Eastment, 2000, OUP).

example. In this example we search for an exact phrase and combination of words. There are a number of reasons for doing this. For a practical example see the modals activity in Chapter 6. This activity is designed to encourage learners to look at the context in which combinations of words are used. A simpler way is to, initially, examine the number of search hits in order to see how common a particular phrase or structure is.[2]

The search engine used in this example is, once again, *Google,* although the method is not exclusive to this search engine. The technique is very simple: the phrase entered in the search box is surrounded by quotation marks. As with all previous searches, the search engine ignores capital letters.

The following are the results of a number of individual searches for functional exponents, the number of hits they produced and a summary of the type of web-pages found:

Figure 1.16
Results of searching for common business exponents using *Google*

Phrase	Hits	Types of web-pages
"i'll put you through"	1,120	Business English web-sites, particularly web-pages featuring telephone language.
"i couldn't agree more"	35,000	Mainly contributions to on-line discussion forums, across a wide range of topics. Examining the context in which the exponent is used would elucidate its meaning.
"i can go along with that"	529	More contributions to discussion forums. Other results show the phrase as part of a direct quote, thus displaying it as a spoken form.
"that's out of the question"	2,430	Also often used as a direct quotation of spoken English. Many of the results provide a context which shows the strength of the exponent.
"the subject of my presentation is"	44	Results are almost exclusively the transcripts of actual presentations.
"let me conclude by saying"	2,130	Again most results are from transcripts of presentations, addresses, briefings, statements, speeches, opening remarks, etc.
"i look forward to"	972,000	Results include the text of letters, advice on writing formal letters, speeches and statements. Many demonstrate the use of the ing form of the verb following this exponent.

This technique and / or the results it produces can be used in a variety of ways before, during and after lessons by teachers and learners alike.

7.4 Other ways of finding things on the Web

Often the best advice to find interesting and useful web-sites and services is to keep your eyes peeled when looking through newspapers, magazines and other web-sites. Many of the most visited and famous web-sites had their existence propagated by word-of-mouth or e-mail. Look through a publication that has advertisements. Most companies put the URL of their corporate web-site somewhere in the copy. Ask your learners for the URL of their company's web-site. Many newspapers have regular, weekly pages or sections devoted to on-line issues and are a goldmine of Web URLs.

Not only has someone else done the searching for you, this has the added value of some sort of description and idea of the usefulness of the web-sites.

As you surf the Web itself, take note of interesting web-sites you visit. You may not have time to investigate there and then. Save the web-page in a logical category in your 'Favourites' and go back and look at it another time.

If you can develop this habit for collecting useful-looking web-site and page URLs, you may find that some of them become old, and highly appreciated friends. You can save a lot of time not having to search for things using the above methods all the time. Also, as you share useful URLs with friends and colleagues they tend to reciprocate, thus allowing everyone's list of 'Favourites' to grow rapidly.

8 Web audio and video

It is possible to listen to Web radio and watch Web TV. However, audio files are much larger than text files and video files are even larger. Even with the latest forms of file compression, you need to be aware of the technical compromises which have been made in order to deliver these files over the Web in some sort of useable form.

In order to take advantage of any type of audio, the computer you are using needs to have a soundcard and some sort of output device such as speakers or headphones. These are standard on home computers but not always on those bought for office use. If you are using audio in-company, it is important to check whether there are rules about playing it on the computers. Indeed a company's firewall may prevent the use of many types of Web audio and video.

Streaming | While there are many places on the Web where you can download small audio files (such as web-sites offering help on pronunciation) and video clips, the majority of audio and video presented as radio and television is delivered via a streaming process. Streaming involves the web-site sending your computer a small packet of compressed audio and video. Once this packet has been downloaded, the software you are using plays it. While it is playing this first packet, your computer is downloading the next packet and so on. The result is a continuous stream of audio or video into your computer. This requires extra software, which we will discuss below. It is also important to remember that this streamed audio and video is not stored on your computer's hard drive. Almost like radio and television coming through the airwaves, it is not automatically retained. Therefore, in order to use these streamed products you need to be connected to the Internet for the length of the audio or video clip. (For more information see Chapter 9.)

Compression | The second thing to be aware of is the results of compression. Compressing audio can still produce reasonably high-quality sound that is as intelligible as the original source material. That is to say, if the original recording was low quality, there is little that can be done to raise this quality. However, the majority of streaming audio available via the Internet that is useful to Business English teachers and learners will have been produced professionally in a radio or recording studio. Compressing video is a different story. Streaming video has to carry not only the moving pictures of the clip but also the sound. The results are two-fold: the quality of the sound on video clips is significantly poorer than that of audio clips and the size of the video picture is tiny. On a monitor with a 17 inch (43 centimetre) screen set to 800 by 600 pixels resolution, the standard Web TV screen is about 2½ by 2 inches (6 by 5 centimetres). As you increase the resolution of the screen this falls even further. The software for viewing streaming video allows you to increase the size of the viewing screen, even up to almost full-screen. However, since this does not cause any increase in picture resolution the resulting image is often indecipherable. Increasing the viewing screen size also forces the computer's processor to work harder. On older computers the result of this will be that you find yourself watching a series of blurry still images that up-date every few seconds rather than moving video.

Connection speed | A final consideration is connection speed. Although the software you use should be set up to match the speed of your Internet connection the same rule always applies: the slower your connection the longer it takes to download data, whatever its nature or format. With slower connection speeds, what tends to happen is that the player gets ahead of the download process. When this happens the play-back stops while the download tries to catch up. If this happens several times during a two or three minute piece of streaming audio, then your connection speed is insufficient. Generally speaking, if your computer has a 33.6k modem, you will just about be able to listen to streaming Web radio. If you have a 56k modem, then radio will be fine but video will stutter as the play-back gets ahead of the download. When this happens the audio may continue but the picture will freeze and skip frames. Digital lines such as ADSL are fast enough for both audio and video in their current common formats on the Web. Future developments in hardware and software will make full-screen streaming video and hi-fi quality sound an everyday feature of the World Wide Web.

8.1 Audio and video software

As mentioned above, listening to Web radio or watching Web television requires additional software called a media player. This comes in the form of programs that can be run independently but, more importantly for this topic, can act as browser plug-ins. Briefly, a plug-in is a program that runs inside another program when it is required. (For more information see Chapter 9.) What this means for Web audio and video is that when you click on a hyperlink that calls up a clip of streamed audio or video, your browser will check to see whether you have an appropriate piece of software to listen to or view this clip. Once that piece of software has been located the computer opens it and uses it to play the clip.

There are a number of programs available, most of which come in at least two versions: a free one, with limited functionality (although sufficient for the purposes outlined above) and one that costs a nominal amount of money (usually around US$30). The most common way to acquire this software is to download it from the Web. We recommend choosing the free version although this may mean scrutinising the software company's web-page very closely to find the appropriate link.

The main pieces of software available are: *RealPlayer* (see: http://www.real.com), *Windows Media Player* (see: http://www.microsoft.com/windows/windowsmedia/download) and Apple *QuickTime Player* (see: http://www.apple.com/quicktime/download).

Figure 1.17
RealPlayer

Figure 1.18
Windows Media Player

At the time of writing there is still no consensus about a standard, so if you are going to engage in extensive Web listening or watching, you would be advised to have all three media players installed on your computer. Different web-site designers and owners have opted to deliver content for different programs. The BBC, for example, require you to have *RealPlayer* to listen to or watch the many clips on their web-site, especially their news site. Other web-sites will require you to have *QuickTime*. A useful arrangement is the one offered by CNN which gives you the option to use either *RealPlayer* or *Windows Media Player*. However, this is not common yet.

One source of impartial information about these and other media players is the *Webmonkey* web-site. See: http://hotwired.lycos.com/webmonkey

8.2 Where to find audio and video on the Web

The best places to look are the web-sites of companies and organisations which produce audio and video content. Many established radio stations and television channels provide links to short clips as well as to live broadcast material. On the BBC's news web-site, for example, you can hear and see clips ranging from around a minute for a single report, five minutes for a world news summary to thirty minutes for the most recent television news bulletin. On another part of the BBC's web-site dedicated to Radio 4, their radio channel whose content is mainly news and discussion, you can listen live to the channel as it is broadcast.

More and more companies are including audio and video clips on their corporate web-sites. These are sometimes part of the company's marketing strategy, offering the opportunity to watch television commercials for the company's products or see and hear senior management talking about the company's systems and values. More often they are recordings of live Webcasts of the company's AGM featuring presentations by the CEO and financial directors.

When you arrive at a web-site that advertises audio and video content look for the tiny icons that have become the shorthand for links to this content. Audio links are represented by a speaker symbol while video links use a video camera symbol.

9 The Internet and CD-ROM

CD-ROM stands for Compact Disc, Read-Only Memory. CD-ROMs used in language teaching typically provide learners with a rich multimedia environment. Apart from text this can include images, photographs, animations, sound (audio-clips) and video. CD-ROMs are fairly commonplace in Business English teaching, frequently being used either as self-study or reference tools.

9.1 Hybrid discs

Note: a DVD-ROM holds more information on the disc than a CD-ROM. Electronic encyclopedias are now available on DVD. Most new computers nowadays are sold with a DVD drive which can also read CD-ROMs

Increasingly, many CD-ROM publishers are making use of the fact that many computer users have a connection to the Internet. A 'hybrid disc' is a CD-ROM whose features include links to web-sites. Clicking on such a link opens the web-page in either the CD-ROM program or using the computer's default web browser. Doing this on a computer with a permanently open Internet connection, i.e. via a network, results in a transition from data provided by the CD-ROM to that from the web-site that is often too smooth to notice.

This feature allows the user to have access to the latest, relevant information and, effectively, keeps the CD-ROM up-to-date. Many electronic reference works have this feature, e.g. the *Cambridge Learner's Dictionary* (CUP).

9.2 Multimedia dictionaries and the Web

The latest CD-ROM dictionaries have a feature which enables the learner to move his / her mouse over the words on a web-page and automatically see a definition of and hear the pronunciation of that word. See the 'QuickSearch' feature on the *Macmillan English Dictionary* (2002, Macmillan).

For more information on CD-ROM in language teaching, see:

http://www.summertown.co.uk

Figure 1.19
Summertown
Publishing web-site

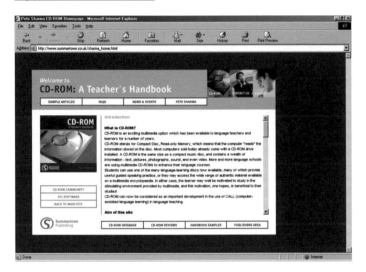

2 The Internet and language teaching

Chapter 2 is divided into the following areas:

1 **Why use the Internet?**
2 **What is on the Internet?**
3 **How to use the Internet**
4 **Incorporating the Internet into the course / lesson**

1 ## Why use the Internet?

The Internet can be used in a number of ways:

- as a source of materials
- as part of a face-to-face language lesson
- as a stimulus for a self-study activity
- as a means of communication between remote users.

Note: As well as the Internet, intranets can be used in a number of ways to help support language learning. Some of the following points also apply to intranets. (See Chapter 1 for an explanation of intranets.)

It may sometimes be enough to simply use the Internet 'because it is there'. In the same way that we can use a text, video clip or audio-cassette to change pace, we may choose to incorporate the Internet into a course or lesson simply for variety. However, it is important to consider exactly why we are choosing to use the Internet. This section looks at why we use the Internet in Business English courses.

1.1 Range of material

The World Wide Web, which resides on the Internet, provides a vast repository of material which can be used in language lessons. Now that the Internet has become an everyday part of life for many, material from the Internet is readily available to many teachers and learners.

Internet search facilities have revolutionised our approach to obtaining materials in Business English. Before their existence, a teacher would frequently encounter difficulties finding a useful or relevant text for a learner in a specific field such as law or accountancy. This has changed, and the Business English teacher can now find something on practically any area to use with learners.

1.2 Authenticity

Authenticity ensures both face-validity and content-validity. Authentic materials such as company brochures, product samples and the learners' own material are already used widely in Business English, as a way of supplementing, say, a published course book, framework or teacher-produced material. Using authentic material is motivating for many learners.

Nevertheless, using authentic material can have its downside. Exposing low level learners or learners who are new to their field of business to authentic texts can be overwhelming. It is important to ensure that the language task set is appropriate in level and achievability for the learners in question.

1.3 Currency

Material taken from the Web can be completely up-to-date e.g. the latest financial figures or a dramatic breaking news story. This currency of information can be exploited; for instance if a major event occurs within a learner's company, competitor's company or field of business, and develops during their language course, it can be incorporated into the lessons. Try the BBC World Service web-site.

Figure 2.1
BBC World Service
web-site

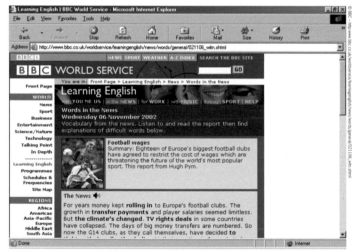

1.4 Expectations

Many Business English learners today expect a language school to have the facilities to access the Internet as part of their teaching programme, either as a way of finding source materials, or as part of the classroom procedure itself.

Many learners will be used to fast connections in their workplace, and may be unforgiving if the connection speed is slow in their lessons.

While participating on intensive courses, learners appreciate the chance to access their private e-mail or company intranet. However, because this usually involves using their own language, it can be distracting to their studies. There may also be cost implications so you may have to consult the school or company you are working for.

1.5 Interactivity

There is often 'added value' in using a web-site, alongside other types of material such as texts, audio-cassettes, videos and CD-ROMs, often due to the fact that a site may be 'interactive'.

This usually means that the user can interact with the material on a web-site in some way, which provides some sort of active response. This is opposed to say reading a printed text, which is primarily linear. Types of interactivity include:

- choice of route through the material. This can be determined by the user, using hyperlinks to move off in different directions
- choice of media. The user can make decisions about how many times to listen to an audio-clip, or choose when and whether to access a video clip etc
- submission of answers or information. The learner fills in a form or completes an activity, submits this and receives some kind of feedback
- customising. The learner can input data and receive specific information. For example, the user can calculate costs of transporting gas across Europe on the Wingas web-site, or customise the design of a car on the Honda web-site. On-line booking of travel tickets and accommodation are further examples of this type of interactivity.

1.6 Developing Internet skills

There is no doubt that using the Internet is central to much business practice. It may well be that the teacher wishes to afford opportunities for their learners to replicate tasks they perform in real-life, in the same way that many business schools offer students a chance to make telephone calls as a way of practising telephone conversations. For example, the learner may be a web-site designer, needing to be able to discuss his / her product with clients. Showing the web-site to the teacher live, therefore, can be motivating and relevant. In this example, the use of the Internet is driven by learner needs. Further practice can include composing e-mails, or even replicating video-conferencing.

A key point to remember is that new types of skills, including language skills, have emerged during the technological revolution. Learners now need to research using the Internet, and to use Web features such as a menu or a search engine. Learners can receive practice in these sub-skills through using the Internet during their language course and consequently develop their Internet skills.

2   What is on the Internet?

The list of what is accessible from the Internet is almost inexhaustible and includes the following.

On the Web:

- company sites, annual reports
- product descriptions, pictures, graphs, diagrams
- financial data
- advertisements and order forms
- multi-lingual dictionaries, specialist dictionaries
- translation tools
- electronic journals and newspapers
- news sites
- academic articles, scientific texts
- country information
- travel information, virtual museum tours
- business English materials
- learning material written for the Web, such as *Market Leader* resources (Longman)
- language activities
- language courses
- sites for Business English language teachers
- information about language schools, application forms
- forums to share information and ideas (learners can post messages to other learners interested in similar subject areas)
- virtual areas such as self-access centres and MOOs.

On the Internet:

- programmes such as e-mail and *NetMeeting* which enable communication across the Internet.

On intranets:

- in-company journals and news bulletins which can be exploited by the teacher and learners working in-company
- messages posted for learners who miss classes
- language tests, timetables and homework tasks.

Figure 2.2
Longman's *Market Leader* resources on the Web

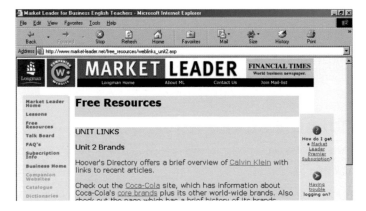

How to use the Internet

There are many varied ways of incorporating the Internet into the teaching of Business English. In this section, we will look at some common ways in which the World Wide Web can be used in language teaching. Specific practical ideas can be found in Part two of this book.

First, a number of distinctions can be made:

1. The Web as a source of material vs the Internet as a means of communication (CMC).

2. A specific task using the Internet integrated into a language lesson vs the Internet used independently by learners, perhaps after a course or during their free time.

3. Studying from a language course available to everyone on the Internet vs using purposely created Web-based materials integrated into a language course, sometimes called 'blended learning'.

Figure 2.3
Using the Internet in language teaching

Independent use	Language lessons	Language courses
1 **Alone and unaided** e.g. learners do an activity on present simple vs present continuous; type in a grammar question to a clinic; look up a word on a Web dictionary; use a translation tool, look up the 'idiom of the day'.	1 **Pre-lesson** e.g. learners research facts and figures on-line before giving a presentation. 2 **During a lesson** e.g. teacher sets an on-line task which learners do in self-access. 3 **Post-lesson** e.g. learners research on-line in order to write a summary. Note: Can also be done off-line.	1 **A free-standing, complete course** e.g. learners do a writing skills course on the Internet. They log on, type in their credit card details and take the course.
2 **Support for another product** e.g. learners subscribe to a service such as *Market Leader* (Longman) which sends regular updates or OUP which offers teaching material / worksheets etc. to enrich their course books.	2 **Projects** e.g. learners use the Internet for long-term projects. A sophisticated project would involve creating a real web-site. This can be very motivating. Often, expertise lies within the group.	2 **A virtual school** e.g. learners never meet the teacher in person, but communicate through a variety of means: via the keyboard; through forums and chat rooms; maybe using a desk-top video-camera. Perhaps the teacher and the learners have access to a shared whiteboard.
3 **After the course** e.g. learners join a discussion forum; listen to the news daily on the BBC or CNN sites etc. Post-course support can be more focussed on the individual's language needs.	3 **Materials** e.g. teacher uses material from the Internet to enrich lessons / give face validity. Materials are easily customisable in word processing packages e.g. gapped / simplified.[1] The use of search engines has radically changed Business English.	3 **Blended learning** e.g. learners participate in a course which comprises a mixture of F2F (face-to-face) and on-line study – learners have lessons and an option to access on-line support. Materials are available on-line.
Comments There's a lot out there! Varying quality. No editing in some cases. This could be an enrichment to something else a learner is doing, such as going to classes or as a follow-up to a course, or just something out of interest.	**Comments** It is always a good idea to try out the sites to be used beforehand, to check they are functioning and to have a back-up activity in case of problems.	**Comments** E-lessons are lessons available on the Internet. They are not necessarily live. In the second example, the technology enables contact with a teacher, but the teacher and learner never meet face-to-face.

[1] See Chapter 9 for information on copyright.

3.1 Using the Internet in the classroom

It is recommendable to transfer usual good pedagogic practice to a lesson using the Internet, such as establishing lesson aims, and ensuring learners are aware of why they are doing a particular task. However, it is even more important to have a back-up activity in mind, in case a technical hitch occurs. For some activities, having a print-out of a particular page may suffice.

There are many ways of teaching with the Internet. This can be done on-line or off-line.

On-line | A typical lesson using the Internet on-line may involve the following stages.
1. Pre-computer work: the teacher engages the learner in some way, setting a task.
2. Computer work: the learner is involved in a task on the Internet, such as searching for data on the Web.
3. Post-computer work: the learner reports back his / her findings.

Other possible teaching scenarios are:
• the learners use a web-site as part of a presentation
• the teacher uses a web-site as part of a presentation
• the learners collaborate on a project, such as the creation of a web-site
• the learners do a task using a web-site before arriving for a lesson (pre-lesson task)
• the learners follow up a language lesson with a specific task (post-lesson task).

Off-line | This may include using materials taken from a web-site in the lesson. It is common to use the Internet to search for a text, which is then used as a stimulus in the lesson. The text can be adapted in some way, for example, copied into a word processor and gapped, or simplified.

A site may have been accessed before the lesson, and the teacher can use the web-pages without actually connecting to the Internet again. This is because the pages used are stored in the computer's cache memory, or saved on the hard drive. (See Chapter 9.)

Another way of accessing a site off-line is to copy the relevant web-pages onto some sort of removable media. This has the advantage that the computer need not have an Internet connection, opening up the possibility of using other computers in the centre for work on that web-site. The disadvantage is that none of the hyperlinks connecting these web-pages will actually be active.

3.2 Using the Internet for self-study

There are many sites expressly created for language learning. They include sites with:
- listening clips
- language explanations
- on-line dictionaries
- on-line activities
- language games
- simulations
- complete courses
- tests.

Some of these sites are pedagogically sound, others are rather poor. For more information on the range of sites for language learning, see *The Internet and ELT* (David Eastment, 1999, Summertown Publishing).

The essence of this use of the Internet is that it can be accessed by any learner at any time, regardless of level and they can use the Internet as they wish in order to benefit their study of language. A Business English teacher may wish to direct learners towards specific sites or activities, however, to ensure good quality, for example in a session at the end of a language course, offering advice on how to continue studying.

3.3 Using the Internet for blended learning

Blended learning refers to the mix of media used in language teaching, between face-to-face and study supported by technology. This is also referred to as POLL, or Partially On-Line Learning. The big advantage of this approach is that it allows learning to take place at a distance as well as in the classroom.

Learners may study from Web-based teaching materials, interspersed with lessons with a teacher. A large teaching organisation, such as Linguarama, may enrich the face-to-face part of a course through offering a Web-based option. A virtual school such the one run by International House Barcelona, creates Web-based material and then supports learners efforts with other forms of CMC such as e-mail.

The advantages of using Web-based materials are many. One key advantage of using materials specifically created for the Web is that they are purposely designed to complement face-to-face learning, and do not attempt to replicate the role of the teacher. An example is *Linguarama Direct*.

Figure 2.4
Linguarama Direct

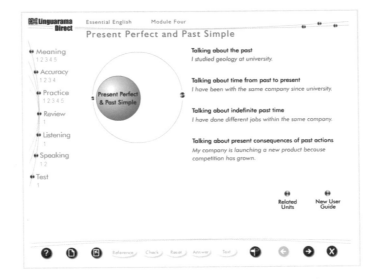

3.4 Using the Internet as a source of materials

The Internet is a rich source of authentic materials. Teachers can find suitable sites and texts and exploit these in the classroom. Here are some possible activities.

Searching | Typically, the business English learner works in a specialised field and few suitable materials (and virtually no published materials) exist in this area. The teacher can run a search and look for a relevant text for use with this learner or the learner him / herself can search for a suitable text to use in a lesson. (See Chapter 1 for advice on effective searching.)

Downloading | The teacher or learner can download material to be used later during a lesson. A typical example would be downloading and saving a PDF file for printing out at a later date. A web-page can be saved or stored on the hard drive or to some form of removable media such as floppy disc, zip cartridge or CD-R for later use.

Note: See Chapter 9 for information on copyright.

Customising text | The teacher can modify the text in some way e.g.:
- change the font size, embolden and italicise
- gap a text and place the missing words in a box on the same activity sheet
- simplify the text in some way to suit the level of the learner, e.g. by deleting obscure words
- incorporate an image or graphic into teacher-produced materials.

Printing | The teacher or learner can print out a page, or section of a page, for use in a lesson. It is easy to highlight a short section of text, copy it to the clipboard and paste into a word-processor in order to print it out.

4 Incorporating the Internet into the course / lesson

Business English teachers work in a wide variety of situations teaching groups or individual learners, typically:

- in-company
- in a language school
- running seminars in a conference centre.

The location and accessibility of the computers available does affect the way in which the teacher can incorporate lessons using the Internet into the teaching programme. (See Figures 2.5 and 2.6 for some possible options.)

Figure 2.5
Layout of the classroom

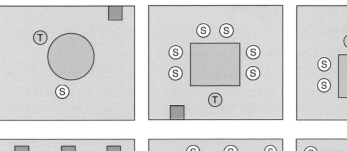

Figure 2.6
Layout of the self-access centre

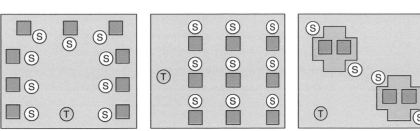

4.1 Computers with Internet access in the classroom

We will now examine some common scenarios.

One-to-one | The teacher and learner work in a classroom with a computer in the room.

The advantages of this situation include:

- constant access is available, allowing the teacher to incorporate activities involving the Internet seamlessly into the lesson, as and when required.
- easy access for both teacher and learner to the screen, to look at sites / work on an activity.

The disadvantages include:

- the teacher may not have had the opportunity to check the web-site is functioning beforehand due to access to the room being restricted.

Note: A beamer (also known as a data projector) is a device for projecting the image of a computer screen onto a wall or screen. Many Business English learners are familiar with using them for giving business presentations. Increasing numbers of language schools are purchasing beamers and owning this piece of hardware opens up a number of possibilities.

Groups | A computer available in the classroom can also be used with a group. The teacher can use a beamer (see Figure 2.7) to present the Internet screen. This is useful for explaining a task which is to be done in the self-access centre or as part of a teacher presentation on the use of language. Alternatively the learners may use a beamer for their own presentations. It is less common to use an LCD (Liquid Crystal Display) tablet which sits on top of an over-head projector, but the same effect can be achieved, that of projecting the computer screen.

Figure 2.7
A beamer

4.2 Computers in the self-access centre

A number of computers are generally available in a self-access centre, either in a language school or in-company. The teacher can take the group or individual learner to this centre for part of a lesson or for a complete lesson, by booking the room.

The advantages of this situation include:
• learners can work at their own pace on an activity
• the teacher has a clear role as a monitor / facilitator
• the teacher can ensure that learners 'stay on task' through monitoring the activity.

The disadvantages include:
• the teacher may have minimum control over the room layout; the layout of the self-access centre is frequently fine for self-study, but actually quite poor for running a lesson
• the use of a firewall in large number of companies, which may prevent the receiving of attachments in activities which involve sending e-mails.

4.3 Learners have Internet access in-company / at home

Learners may have access to the Internet at work or at home. Learners with a lap-top can also dedicate committed time to studying even while travelling on business. This access allows the teacher to set pre- and post-lesson tasks.

The advantages of this situation include:
• enriching and relevant tasks can be set outside course time
• more controlled aspects of teaching, such as certain types of pronunciation practice, consolidation of structure through repetition etc. can be done outside the face-to-face parts of the course.

The disadvantages include:
• the learner may be tempted to surf aimlessly and move off-task
• the lack of motivation to work on self-study tasks.

3 The Internet and Business English teaching

Chapter 3 is divided into the following areas:

1 **Language areas**
2 **Business areas**

1 ▢ Language areas

This section will first examine the four language skills (speaking, listening, reading and writing) discretely, explore how the Internet can help with integrated skills work, and then look at other aspects of language teaching: grammar, vocabulary and phonology.

1.1 Speaking

The teacher usually plays a specific role in fluency practice – that of a communication partner, providing a stimulus, reacting to learner utterances etc. The Internet, however, can be the 'means' through which speaking takes place, such as desk-top video-conferencing or for making telephone calls. Additionally, the Web itself can be useful in providing speaking practice in a number of ways.

The Web can provide an environment facilitating discussion and thereby, it is hoped, promoting learning. For instance, two learners can work together at the computer screen discussing possible answers to a grammar activity. The speaking which the learners do around the activity can provide useful practice for language development. It will probably include:

- negotiation of meaning
- asking for clarification
- listening practice
- fluency practice.

Such discussions are usually monitored by the teacher and feedback given.

The computer can also provide data to be used in a simulation (see later in this chapter for details). For example, calculations can be performed (e.g. finding the optimum route between two places) and the data used to stimulate and support the discussion.

The list of topic areas for discussion within Business English is inexhaustible, and the Web can provide a wide range of materials to be used as a springboard. Here is a selection of areas typically discussed with Business English learners, including a list of sites where useful information can be found.

Figure 3.1
Discussion topic
areas and possible
web-sites

Topic area	Web-site
Decision-making	http://www.mindtools.com
Globalisation	http://www.globalisation.com/ http://www.worldexploitation.com
Management styles	http://www.mapnp.org/library/mgmnt/cntmpory.htm
Motivation	http://www.accel-team.com
Quality control	http://www.goalqpc.com/whatweteach/tqmwheel.asp
Time management	http://www.mindtools.com
Team building	http://www.teamtechnology.co.uk/tt/t-articl/tb-basic.htm

1.2 Listening

There are a number of benefits in using the Web for listening practice:

- the learner can be in charge of the controls. The BBC News site uses *RealPlayer* which features a slider allowing the learner to move through the clip, and repeat certain sections at will. (See Chapter 1 for more information about the correct plug-in.)
- the learner can listen to a stretch of language and read the transcript or a related text, and so re-enforce what he / she has listened to earlier.
- a listening clip can be more up-to-date than a published audio cassette, video or CD-ROM. Today's news may include a relevant business topic, such as a takeover.
- the learner can access the listening task at any time in his / her office or at home, encouraging learner autonomy.
- listening on the Internet can provide exposure to native speakers, offering a range of accents, phonological features of language such as weak forms and sounds in connected speech.

Figure 3.2
RealPlayer

While CMC activities naturally lend themselves to developing listening skills, the Web provides a range of many different types of listening in both audio and video clips, including:

- discrete items (e.g. audio clips of phonemes)
- short stretches of language (e.g. sample sentences for intonation practice)
- longer discourse (e.g. business presentations, the news).

However, some material may be unsuitable. Clips from films on one EFL site, for example, are full of bad language. Just because it is there does not mean it is suitable.

Moreover, listeners may encounter problems such as:
- slow download times for an audio clip
- the need to choose which software to download (*RealPlayer* and / or *Windows Media Player*)
- encountering messages asking if you wish to download software updates.

Figure 3.3
BBC World Service
news web-site

1.3 Reading

Reading on-screen is different to reading on paper. There are a number of reasons for this – such as the readability of the font type and the size of the screen. However, computers with Internet access used for language teaching are often connected to a printer, so learners can print out the text and read it as a hard copy should they so wish.

Hyper-text | The existence of a hyper-linked Web environment has been changing the way we access information, or at least offering alternatives to how we would usually do this in a linear text. There are many differences in reading a piece of text on the Internet as opposed to a paper-based document, especially if the text is set in a multimedia context allowing access to sound or additional information. These differences include:
- the use of annotations: the learner can access additional information as and when they need it
- the use of hyper-links: the learner can surf sideways and explore at will.

Incidental reading | Incidental reading occurs on the Web in all sorts of ways. The learner does not only read the chosen text but also reads:

- instructions
- menus
- help files.

Deduction from context occurs when learners extrapolate the meaning of the *Windows'* instructions or icon labels because they may already know them in their native language.

Meaning on demand | The 'QuickSearch' mode on the *Macmillan English Dictionary* on CD-ROM (2002) enables users to move their mouse over a word and see the meaning appear in the on-screen window. Learners can also hear the word pronounced at the same time.

Figure 3.4
*QuickSearch on
Macmillan English
Dictionary* (2002)

The same way of accessing meaning from a web-page is possible by using the *Oxford Advanced Genie* (OUP).

Figure 3.5
*Oxford Advanced
Genie*

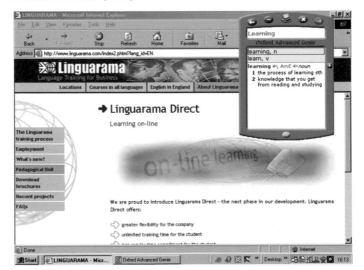

1.4 Writing

There is a natural connection between learners using the keyboard and working on the skill of writing, and performing Internet-based tasks.

When using the Internet, writing can be synchronous, e.g. using chat, or asynchronous, e.g. sending e-mails. When using the Web, there are a number of instances when learners practise asynchronous writing such as:

- completing forms
- filling in an on-line questionnaire or C.V.
- sending a message to a bulletin board or forum.

For more on writing, see Section 2.5 below.

1.5 Integrated skills

Learners do not necessarily practise the language skills in isolation when using the Web, but frequently do integrated skills work. Listening can be easily combined with another skill e.g. learners can listen and read together. This can be similar to the activities found on many CD-ROMs. Indeed, the advantages of using multimedia are well-documented.

Examples of integrated skills work are:

- learners listen to a clip and read the text at the same time (listening + reading)
- learners fill in an application form and submit it (reading + writing)
- learners engage in an e-mail project (reading + writing)
- learners listen to a phonology activity and repeat the sounds (listening + speaking)
- learners take an on-line lesson (all four skills).

Of course, one of the advantages of using multimedia is that the learner can be placed in control of the amount and type of exposure to language they receive, and that it will probably involve integrated skills work.

1.6 Grammar

The Internet affords many opportunities for the study of grammar. There are countless sites which offer practice in discrete items of grammar, such as work on tenses, prepositions and conditional forms.

Figure 3.6
englishclub.com

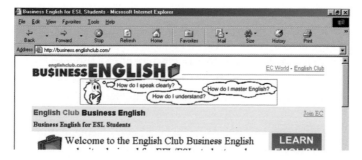

Grammar practice activities | For a site which is typical of examples of grammar activities on the Web, see: http://ccc.comnet.edu/grammar/ or http://englishclub.com/

Many such activities are free-standing. Others are part of a suite of activities provided by a school. For an example of the type of grammar support available in a virtual school, see: http://www.netlanguages.com

Grammar clinic | A learner can e-mail a question to a grammar clinic such as: http://www.lydbury.co.uk/discus/index.html

The learner can post a message about a grammatical area on the bulletin board, or contact another user by e-mail.

Figure 3.7
Lydbury Grammar
Clinic

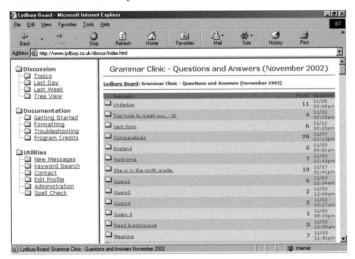

1.7 Vocabulary

Note: A good deal of the English on the Internet is US English, which means the vocabulary and spelling is US-oriented. Additionally, there are many spelling mistakes on the Web and other areas of Internet usage, either because the material is unedited, or because it resides in a 'forum' where users include non-native speakers, people typing quickly or even levels of native-speaker illiteracy.

The Internet provides a rich environment for exposure to lexis.

Specialised vocabulary | There are many specialist dictionaries on the Internet, including medical English, legal English, financial English etc.

For a list of specialist dictionaries, see: http://www.lib.uwaterloo.ca/libguides/1-10.html or http://www.sciencekomm.at/advice/dict.html

Translation tools | A high number of one-to-one relationships between words in English and other languages is a feature of ESP (English for Special Purposes). In this area, translation sites can be useful. Try: http://europa.eu.int/eurodicautom/login.jsp

There are translation services available on the Web. Exploring the usefulness of sites such as: http://uk.altavista.com/babelfish is a possible learner activity.

Of course, the area of translation is not unproblematic, and many teachers would urge their learners to check translations in a mono-lingual dictionary.

Collocation | Collocation is an important feature of Business English. The Web is full of texts which are relevant to particular learners and the nature of ESP means that such texts tend to be a rich source of useful collocations. (See: Chapter 6: *Collocation search*.)

Learners can gain valuable insights into this area through sites such as:
http://titania.cobuild.collins.co.uk/form.html

There are also activities on collocation, such as those at:
http://esl.about.com/library/business/be_vocab_buscollocationsl.htm

On-line dictionaries | The explosion of technology has produced a huge number of neologisms, and of course the Web is the ideal place to look up the meanings of such terms. An on-line dictionary can be up-to-date to the minute. Try:
http://www.longman.com

Additionally, many CD-ROMs allow access to the Internet. Users of the *Cambridge Learner's Dictionary* on CD-ROM have hyper-linked access to a range of CUP dictionaries on the Web.

Figure 3.8
Cambridge Learner's Dictionary

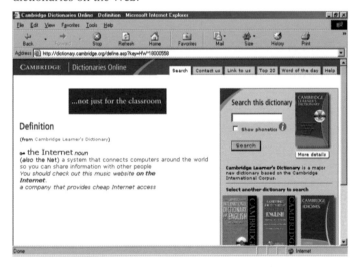

Annotations | The use of annotations is one of the reasons why the multimedia environment of the Web is useful in the area of lexis. It gives learners the chance to access meaning on demand. Some language teaching web-sites for instance are actually linked to on-line dictionaries, allowing learners to access the meaning of a word or phrase simply by clicking on it.

1.8 Phonology

There are a number of ways in which the Web can help learners in the area of phonology.

Phonemes | Learners can obtain downloadable audio files of the individual phonemes in British Received Pronunciation or in American English. They can access a wealth of information on phonology covering areas such as accent and the phonemic script, or they can obtain phonetic fonts for use on word processors. A good place to start is: http://www2.arts.gla.ac.uk/IPA/ipa.html

Word stress | Learners can click on parts of a word and hear how to stress each syllable. See: http://pronunciation.englishclub.com/word-stress4.htm

2 Business areas

In this section, we will firstly look at using the World Wide Web to help in the teaching of some core business skills (giving presentations, attending meetings, negotiating, telephoning, writing). Then, we will look at a range of elements which are features of many Business English courses (ESP, simulations, games, inter-cultural training, tests, examinations).

2.1 Presentations

Note: Some companies place a presentation on the company intranet. This can then be accessed by the learner and practise giving the presentation with his / her teacher in an in-company lesson. If the presentation is on the company extranet, then the learner can access it from any computer, and not only a computer in the company offices.

Giving presentations | Learners can access information on giving presentations on the Web. For instance they can find:
- useful phrases for giving presentations
- tips for giving successful presentations
- model business presentations given by native speakers.

Try: http://www.research.ucla.edu/era/present/

Learners can also use the vast resources offered by the Web in preparing their presentations, looking for information, selecting and incorporating text, and even including photographs and graphics. The text and graphics can be photocopied onto OHP transparencies if learners are not making use of e.g. *PowerPoint*.

Presentation software | Many Business English learners need to give presentations using Microsoft *PowerPoint* or Corel *Presentation* software. Sometimes the creation of the slide-show is done by a design company, but quite often learners prepare their own slide-show. They can use the Web in their preparation of their presentations.

Learners can:
- read professional tips for creating presentations in *PowerPoint*
- cut and paste text from the Web and insert it into their presentation
- import graphics and photographs taken from the Web
- create a hyper-link to a web-site which can be accessed during the presentation.

2.2 Meetings

Learners can access information on successful performance in business meetings on the Web. Learners can:

- listen to discrete phrases which they can use in meetings, such as 'Let's get down to business, shall we?'
- work their way through a business maze on the language of meetings.

Try: http://ww.celt.stir.ac.uk/staff/HIGDOX/VALLANCE/Diss/FP.HTM

The site allows learners to work through the decision maze and thereby 'practise' language skills such as: starting meetings, presenting opinions, accepting and rejecting proposals.

2.3 Negotiations

Learners can access information on the business skill of negotiations on the Web. For instance, learners can access:
- useful phrases for negotiating
- useful hints on the business skill of negotiating.

Try: http://www.negotiationskills.com

John Mole, the author of *Mind your Manners* (1996, Nicholas Brealey Publishing), has a site worth investigating at: http://www.johnmole.com

It includes the 'DEAL' map, describing power and relationship in negotiating; tactics – how to build a negotiation; information on negotiating gambits and how to counter them.

2.4 Telephoning

Learners can:
- look up communication skills hints, such as 'Tips for getting people to slow down'
- access useful expressions for using on the telephone.

Try: http://esl.about.com/library/speaking/bltelephone_tips.htm

Teachers can find:
- ideas for running telephone role plays
- activities for helping their learners with speaking practice.

2.5 Business writing

Many Business English learners typically need to write business letters, e-mails, and reports.

Letter writing | Learners can access:
- letter-writing tips
- information on punctuation, style and layout
- sample letters.

Web-sites help learners through writing tasks such as composing a letter of complaint and can supply a salutation, heading and closure. Try:
http://www.businessletterpunch.com

E-mail | Learners can find out about e-mail conventions at: http://www.email.txt

For e-mail tips and techniques, see: http://www.webfoot.com/advice/email.top.html

Learners can access useful information on:

- the use of 'emoticons' (a mixture of 'emotion' and 'icon' used to indicate such ideas as 'I'm making a joke' which are difficult to transmit in e-mails)
- e-mail conventions e.g. CAPITALISATION means you are shouting; reply within 24 hours etc.

Report writing | Many learners need to write reports in English. Typically the reports are, for example, audit, technical, project and quality assurance reports.

The Web can also be a source of:

- information on the lay-out of reports
- standard report templates.

2.6 ESP

'Search engines were designed for the need and purposes of ESP teachers. Whatever the vocabulary area, you are sure to find a related text on the Internet. Once you've found a text, copy and paste it into a document and use it to create language activities' (John Hughes, 'English for rocket scientists' in *English Teaching Professional*, Issue 21, October 2001).

As we have seen earlier, the Internet has radically affected our approach to gathering content-specific materials in specialist areas. Materials taken from the Internet can be relevant and motivating; have face-validity; and have the advantages of authenticity.

Conversely, they can date quickly. The teacher needs to consider the type of task they employ while using such authentic materials, and issues such as learner overload.

Teachers can suggest relevant sites to learners in a session at the end of a course, suggesting ways of continuing to study. These could include professional forums, on-line journals and sources of articles. (See Chapter 8 for a list of ESP areas and suitable sites to recommend).

2.7 Simulations

Simulations can be done using the Internet as a delivery tool. A major bonus is the opportunity to run simulations where the participants are in different locations.

There are many factors running such projects which have to be taken into consideration, such as regional time differences between participants and the use of firewalls. There are documented examples of projects on the Internet.

For more information on a wide variety of projects, see: http://www.hut.fi/~rvilmi

2.8 Business English games

Playing games on the Web is something many learners may choose to do anyway. Some Business English learners may find playing such games motivating and relaxing, an affective factor noted in language acquisition studies.

Typical games on the Internet which we have found useful for language teaching purposes include:

- Hangman at: http://games.englishclub.com/hangman/
- Business crosswords at: http://www.better-english.com/crosswords/test.htm

2.9 Inter-cultural training

Inter-cultural training, sometimes called Cultural Awareness Training (CAT), is well-established as an important aspect of Business English. It is generally accepted that it is difficult, and occasionally impossible, to do business successfully with other cultures without a certain level of knowledge about that culture. This type of training can be divided into cultural awareness training, focusing on a specific culture, and cross-cultural training, such as the study of critical incidents at the interface between two cultures.

The information provided on the Web includes areas such as taboos and a wealth of cultural information such as food, customs etc. Typical sites offer 'the cultural knowledge needed to travel, conduct business and live overseas' in much the same way that published material offers such information. Try: http://www.countryreports.org/

Figure 3.9
*Country
Reports.org™*

Trainers competent in running this type of course sometimes suggest a web-site that participants can subscribe to, such as: http://www.dialogin.com/

2.10 Language testing

The opportunity for a learner to find out their level of English by taking a test on the Internet is a common claim made by many web-sites. Such claims should be approached with caution. It is easier to set and mark certain test items electronically. These items may therefore be given precedence over other components, such as setting a creative writing composition as part of the work to be evaluated. Despite advances in the computerised marking of compositions, marking such a test would probably require a human moderator.

Some of the traditional face-to-face elements of a test can be done on-line. However, very much like a telephone test, it may not provide the same result as a personal interview due to factors such as nervousness, and dealing with the elements of technology involved. Learners can, for instance, write a composition and e-mail it to the teacher, or record a short piece of speech on the computer, save it in a common audio file format and e-mail this to the teacher as an attachment.

Look at the tests available at: http://www.englishlive.co.uk

Here, a 10 minute test 'checks' (their inverted commas) your reading ability; a 20 minute test 'checks' your reading and listening skills; and a 40 minute test 'checks' your reading and listening skills.

BULATS (Business Language Testing Service) is a practical test of language ability relevant for people working in a business context. See: http://www.bulats.org

Longman English Assessment is a web-based test which offers to test proficiency in reading, listening and written structures. For information, see: http://www.EnglishSuccess.com

2.11 Business English examinations

The Web can be a helpful source of information about Business English examinations, for both learners and teachers.

Figure 3.10
Business English
examination web-sites

Examination	Web-site
Cambridge examinations	http://www.cambridge-efl.org.uk/exam/business/log_bec.htm http://www.cambridge-efl.org.uk/exam
LCCI (London Chamber of Commerce and Industry) examinations	http://www.lccieb.org.uk
General information on examinations	http://www.eflweb.com

2.12 Other

There are other uses of the Web which may be of interest to Business English learners, such as services which provide a daily idiom or business cartoon. See: http://www.tedgoff.com or http://www.glasbergen.com

4 The Internet and the Business English teacher

Chapter 4 is divided into the following areas:

1 **The business world**
2 **Business English materials and courses**
3 **On-line journals and newsletters**
4 **Forums and communities**
5 **Teacher training courses**
6 **On-line conferences**
7 **Employment**
8 **Business English providers**

The aim of this chapter is to give an overview of the professional support available on the Internet for the Business English teacher.

1 The business world

Traditional ways for Business English teachers to keep in touch with what is happening in the world of business include reading the *Financial Times*, magazines such as *The Economist*, and watching business news programmes. The Internet provides another source of information. News sites are frequently updated, and teachers can subscribe to an e-mail list to keep them informed on a particular area.

Web-sites which provide good general coverage of events in the business world are:
http://www.reuters.com
http://www.USAtodaycom
http://www.moneydaily.com
http://www.FT.com
http://www.bloomberg.com

There are also many sites for keeping up-to-date with a specialist area such as personnel or engineering. (See Chapter 8 for further details.)

2 Business English materials and courses

A lot of teaching material is available on the Internet.

2.1 Worksheets

A range of activities and downloadable Business English worksheets can be found on the Web. Try: http://www.better-english.com/exerciselist.html

Some of these are in PDF format and can be downloaded, saved and then printed out. Others are designed to be used on-line.

2.2 Supplementary web-based materials for course books

Some Business English course books have an on-line supplement. Try:
http://www.market-leader.net

The photocopiable material includes pre-reading tasks, comprehension exercises, vocabulary activities and ideas for discussion lessons.

2.3 On-line Business English courses

There is a proliferation of on-line materials for distance learning which can be used by teachers as part of a course, or as pure self-study material. An example of such material is *Talking Business* (Longman). See:
http://www.longman-elt.com/business/teachers/index.html

Materials include ready-prepared Internet activities and resources for the Business English teacher.

There are some Business English courses which can also be done entirely on-line. Here is an example:

Figure 4.1
On-line learning material which includes listening and practice activities taken from http://www.english.is.it

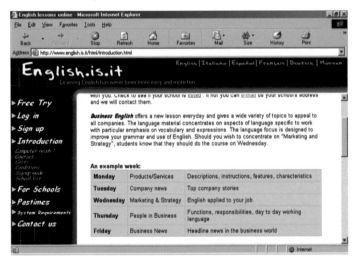

3   On-line journals and newsletters

There are a number of on-line journals and newsletters which are of interest to Business English teachers or learners.

Of interest to learners for instance is *The Language Key* which is a monthly self-study Business English magazine, and is available in an intranet version from:
http://www.languagekey.com Users need *RealPlayer* (see Chapter 1) to hear the audio clips.

Journals of interest to Business English teachers include:

http://www.eltnewsletter.com

This is a weekly newsletter from NetLearn Languages, and is distributed by e-mail every week. The newsletters are archived, a useful feature of on-line journals.

Figure 4.2
ELT Newsletter

Other magazines of interest include:

Spotlight. See: http://www.spoltlight-online.de/

The Internet TESL Journal. See: http://iteslj.org

See also: http://www.4teachers.org

The latter is a Web environment to assist teachers integrating technology into instruction and contains articles, research papers, lesson plans, classroom handouts, teaching ideas and links.

Figure 4.3
Spotlight On-line

Figure 4.4
The Internet TESL Journal

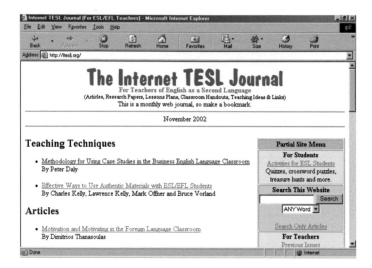

The advantage of such journals and newsletters is that they usually have searchable archives, which are always available. These contain previous editions. The Web is consequently a source of articles on language related areas such as task-based learning; multiple intelligences theory; neuro-linguistic programming.

4 Forums and communities

One of the most well-known forums for Business English teachers is the BESIG (Business English Special Interest Group) forum, details of which can be accessed through the IATEFL (International Association of Teaching English as a Foreign Language) web-site at: http://www.iatefl.org

Joining the BESIG Forum results in you being placed on a listserve. This means that you receive e-mail messages on the forum. The same messages are sent to everyone. You can either read them, or simply delete them. You can either reply directly to the sender, or post a message to everyone in the community.

Typical messages include requests for information about course books, or obtaining ESP materials.

A well-known forum for teacher development, the ELT Forum run by Longman, contains a wealth of useful articles, covering such topics as methodology, language skills and phonology. See: http://www.eltforum.com/

Figure 4.5
BESIG Events Page

Figure 4.6
Possible message on
the BESIG Forum

5
Teacher training courses

Some training courses can be done on-line, such as the COLTE (Certificate in the On-Line Teaching of English) run by NetLearn Languages. The course covers both synchronous ('live') and asynchronous ('not live') methods of teaching. See: http://www.netlearnlanguages.com.

6
On-line conferences

There is a growing trend towards on-line conferences such as the one organised by NetLearn languages, the ELTOC conference (English Language Teaching On-line conference). Session details are usually e-mailed to everyone in a discussion group. The sessions are recorded and archived. See: http://www.ELTOC.com

7   Employment

Details of schools are readily available on their web-sites, and the existence of the Internet has enabled freelance teachers to forward their CV's electronically to prospective employers. Today's job marketplace is certainly global.

For lists of vacancies for Business English teachers, see: http://www.jobs.edufind.com/

8   Business English providers

An on-line presence is essential nowadays for marketing purposes and schools as well as individuals have homepages. Agents looking for courses on the Web use search engines, and prospective clients can sometimes take a virtual tour of a school before choosing. Bookings are frequently done on-line, and safe payment methods allow for on-line course payments.

The Internet can also be used for pedagogical purposes. A school can send out a pre-course task by e-mail and do an on-line needs analysis. Supervised use of on-line resources is possible, as well as on-line tutoring during the course. Schools can then send a post-course task, continue mentoring after the course, and create a community of ex-students, all on-line.

Business English providers include:
E4B English for Business: http://www.e4b.de
Linguarama: http://www.linguarama.com
Lydbury English Centre: http://www.lydbury.co.uk
York Associates: http://www.york-associates.co.uk/intro.htm

Figure 4.7
York Associates

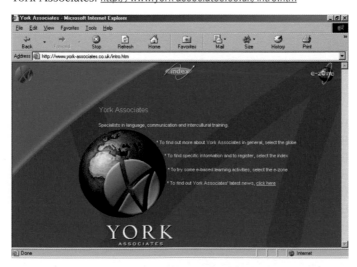

Part two
Practical ideas

5 Learner training

Chapter 5 is divided into the following areas:
1 **Learner training for the Web**
2 **Learner training activities**

1 ▬▬▬ Learner training for the Web

Before using the Web as part of a language lesson or a course, the teacher may feel it is necessary to provide learner training to enable the learners to use it effectively in their language training. This could form part of a wider 'Learning to Learn' component.

'Learning to Learn' is a key part of Business English teaching. During such sessions, learners may focus on learning styles, modalities and related themes; be encouraged to gain insights into their own learning styles; be given study tips and suitable suggestions to develop learning strategies / independent learning.

Learner training can cover one or more areas, depending on the learners' abilities and needs. The range of variables includes the learners':
• professional needs in the area of technology
• attitude towards using technology
• language level
• knowledge about technology
• competence in using technology
• knowledge about using new media in language learning.

Note: Learners with a high level of skill in using the Internet itself, for personal or business reasons, may not have much experience of using the Internet in language teaching.

It is useful to first consider the levels of competence in the technology of both teacher and learner. There are four possible scenarios:

Figure 5.1
Competence level in using technology

HIGH	Teacher and learner		Teacher	Learner
LOW		Teacher and learner	Learner	Teacher

The following activities are particularly useful for the situation where a teacher is relatively new to technology, working with inexperienced learners.

1.1 Learner needs in the area of technology

Each learner or group of learners will use technology in the work-place in different ways. Part of the needs analysis process may be used to gather information about learners' professional needs in the area of technology. To this end, a technology needs analysis, in addition to a traditional language needs analysis, will be useful. (See: Activity 1.)

1.2 Learner attitudes in the area of technology

Learners may have positive or negative attitudes towards technology in general, or using technology in their language course. Such attitudes may well emerge during the needs analysis process. This information will be helpful in gauging the number of Internet-based lessons to include during the course, if any.

1.3 Language level

Note: We suggest putting the vocabulary items onto labels, printing them out and sticking them on cards. This allows for hands-on matching activities.

Elementary learners will certainly need vocabulary input relating to the Internet. This can be done in a number of traditional ways such as presenting vocabulary in a text, building up a lexical field etc. Some of the activities in Section 2 of this chapter provide some support material in this area. (See: Activities 2 – 5.)

1.4 Learner knowledge about technology

For learners with a low level of knowledge about technology itself, pre-course activities could include a simple questionnaire (see: Activity 6) or an exercise matching technical lexis with definitions (see: Activity 7).

1.5 Learner competence in using technology

It can be argued that our role as Business English teachers does not include explicitly teaching Internet skills. However, we frequently raise awareness of such business skills as giving presentations or negotiating. Hence, raising awareness of Internet skills such as searching is certainly becoming increasingly a part of our teaching role. A 'learning by doing' approach means that as learners participate in Internet activities on language courses, they will incidentally improve such skills.

1.6 Using new media in language learning

Note: It is uncommon nowadays to find learners completely new to using the Internet. For learners who are relatively unfamiliar with using the technology, useful skills include:
- using search engines
- researching using hyperlinks
- scanning a web-page for essential information
- saving web-pages
- using drop-down menus.

Learning to Learn | The teacher may wish to show learners the benefits of working in a self-access centre while taking part in a face-to-face course. Such a session may introduce the learner to all aspects of working in the self-access centre, such as books and worksheets, or to other available technologies such as CD-ROM, tape-recorders or satellite television.

An introduction to using the Web to support learning could also present ideas on how to improve language skills such as listening or reading, and using specific sites such as the learners' own company web-site. (See: Activities 8 – 13.)

Continuing to Learn | At the end of a course, there are a range of activities which a learner could usefully do in order to ensure that they remain in contact with the language. These include the regular use of news sites; using a web translator; using a specialised dictionary; joining a community which coincides with a personal or professional interest. These are all part of 'Continuing to Learn'. (See: Activity 14.)

2　Learner training activities

This section contains a number of practical learner training activities. Teachers can select or adapt those most appropriate to each specific individual or group.

Activities

1 Technology needs analysis
2 Internet vocabulary
3 Internet terms
4 Internet collocations
5 How do I say it?
6 Using the Internet in language learning
7 Internet definitions
8 Listening on the Internet
9 Reading web-pages
10 Company web-site evaluation
11 Web-site familiarisation
12 Dictionary comparisons
13 Learner modalities
14 Continuing to Learn

1 Technology needs analysis

Aim	To find out about the learners' experience of using the Internet, attitude towards technology and their needs in this area for during the course.
Procedure	Complete this sheet as an addition to the face-to-face needs analysis you do with your learners.
Variation	The worksheet can be issued and completed by the learners.
Comment	The results should help you decide on the number of lessons which incorporate the Internet in some way.

Technology needs analysis

1. I use the Internet at work:

   regularly ☐

   infrequently ☐

   never ☐

2. I use the Internet in English at work for the following:

   (e.g. e-mail / research / e-commerce / NetMeeting.)

   Details:

3. Other useful information about technology in my company:

   (e.g. connection type / company intranet / accessibility.)

4. Aspects of using the Internet in which I would like to improve:

   (e.g. understanding web-pages in English.)

---

2 Internet vocabulary

Aim To input core vocabulary needed to talk about the Internet.

Level Beginner

Procedure 1 Allow the learners to match the twelve cards into six pairs. Encourage learners who know the terms to explain them to those who are not sure.

E-mail address	teacher@school.com
Company web-site	www.linguarama.com
Homepage	www.school.co.uk — **Homepage** — *Welcome to the school web-site*
Search engine	Google
Web-site address	www.bbc.co.uk
Hyperlink	Click here

2 Invite the learners to ask each other questions using the terms, e.g. what's your e-mail address? Does your company have a web-site?

3 Internet terms

Aim	To familiarise learners with the meta-language needed to use the Internet.
Procedure	Allow the learners to divide the cards into three groups: words they know, words they do not know and those they think they know but wish to check. In a group, higher level learners will explain the terms to the others.

Internet terms (elementary)

homepage	web-site address	search engine	menu
e-mail	surf	favourites	history
attachment	ISP	modem	broadband

© Summertown Publishing 2003. This material is downloadable from http://www.summertown.co.uk

Internet terms (intermediate)

drop-down menu	forum	cache	intranet
extranet	spam	on-line	off-line
newsgroup	chat room	cookie	ADSL

© Summertown Publishing 2003. This material is downloadable from http://www.summertown.co.uk

Comment	This activity works well as a warmer before doing a web-based activity.

4 Internet collocations

Aim   To familiarise learners with the meta-language needed to use the Internet.

Level   Elementary

Procedure   Allow the learners to match the cards. Higher level learners will explain the terms to the others.

Type in	the web-site address
Click on	the hyperlink
Attach	a document
Send	an e-mail
Search	the Web
Add	to favourites

© Summertown Publishing 2003. This material is downloadable from http://www.summertown.co.uk

Comment   This activity works well as a warmer before doing a web-based activity.

5 How do I say it?

Aim   To clarify how we say parts of a web-site or e-mail address

Level   Pre-intermediate

Procedure   Issue the cards to one of the learners in the group. The learner reads them out to the others who copy down the information. The teacher intervenes as necessary.

.com	www.	www.company_	www.company-
teacher@centre.com	www.company/	http://	co.uk

© Summertown Publishing 2003. This material is downloadable from http://www.summertown.co.uk

Comment   This activity works well as a warmer before doing a web-based activity.

6 Using the Internet in language learning

Aim
To provide learners with the opportunity to share experiences and ideas about using the Internet in language teaching.

Level
Pre-intermediate and above

Procedure
Issue the questionnaire and allow the learners to fill in their answers. They can discuss the questions in pairs or small groups before opening up the discussion to the whole class.

Comment
This activity works well to gauge the level of interest in doing web-based activities during the course. This can be quite a confusing, controversial and emotive area, and teachers will need to be sensitive to the possible range of conflicting views from their learners.

Using the Internet in language learning

1. Write a definition of the Internet.

_____

2. What is e-learning?

_____

3. Have you ever studied using the Internet?

   If so, what was your experience?

_____

4. What are the pros of learning a language on the Internet?

_____

5. What are the cons?

_____

6. Have you experienced any aspects of language teaching on the Internet?

   If so, what were your experiences?

---

7 Internet definitions

Aim	To familiarise learners with the meta-language needed to use the Internet.
Level	Intermediate
Procedure	Allow the learners to match the cards. Higher level learners will explain the terms to the others.

Intranet	A private network limited to the internal network of a particular company. Not accessible to outside users via the Internet.
Extranet	A private network which exists as part of the Web. It is protected from general access by a password.
Browser	A computer program used for viewing web-pages downloaded from the World Wide Web. E.g. Microsoft *Internet Explorer*, *Netscape*.
Cookie	A tiny file left on a computer's hard drive by a web-page. When you return to that web-page later the computer will look for it.
Emoticon	Symbols used to represent the feeling or emotions of the person writing an e-mail.
Firewall	A software system to protect a network from outside users accessing the information on the network.

Comment	This activity works well as a warmer before doing a web-based activity.

8 Listening on the Internet

Aim

To introduce learners to the benefits of using the Internet to improve their listening skills.

Procedure

1 **Intro:** Ask these questions:

- Why is listening sometimes difficult? (e.g. speed, accent, weak forms.)

- What different kinds of listening do you know? (e.g. listening for gist, listening for specific information.)

- Explain the advantages of listening on the Web, (e.g. a user can use the on-screen control buttons to pause the recording, and replay a section at will; the recording is digital, so the user can return to a specific point in the recording). Mention that the audio clip can sometimes jump slightly, as it is delivered across the Internet; it is therefore sometimes not as perfect as a studio-recorded language audio-cassette.

2 **Task:** Go to a Web listening site, such as http://www.economist.tv or the BBC site, which has specially prepared listening exercises with texts, at: http://www.bbc.co.uk/worldservice/learningenglish/news/index.shtml

The learners explore the listening facility. (The teacher may need to facilitate.)

3 **Post-task:** The learners report back on the ways in which the web-site could help them develop their listening skills.

Comment

It is recommendable to use head-phones. This activity can be done using specific worksheets such as those in Chapter 6.

9 Reading web-pages

Aim

To show learners the benefits of using a CD-ROM dictionary in order to access meaning when reading a web-page.

Assumptions

The teacher is familiar with the *Oxford Genie* (OUP) or 'Quicksearch' on the *Macmillan English Dictionary* CD-ROM (2002, Macmillan).

Procedure

1 **Intro:** Ask the learners these questions:

- What is the difference between skimming and scanning? (Reading for gist vs. reading for specific information.)

- What reading strategies do you use in order to understand an authentic text? (e.g. predict the content; guess the meaning of a word from context; ask questions about the text before reading.)

2 **Pre-task:** Demonstrate the use of 'QuickSearch' or the *Oxford Genie* as an aid to reading web-pages. (Point out that this can also be used to read e-mails, or as a way of checking spelling when writing in *Word* or *PowerPoint*.)

3 **Task:** The learners have hands-on practice.

10 Company web-site evaluation

To exploit learners' company web-sites to practise fluency and vocabulary.

Procedure

1 **Pre-lesson:** It is vital that the teacher checks if such a site exists, and if it does that there is an English version. Printing out a page or two is a useful back-up to have for this lesson.

2 **Intro:** Ask the learners if they have visited their company web-site. If so, what is it like?

3 **Pre-task:** Issue the 'Company web-site evaluation sheet'. Explain each category:
   • Site description: size / features / any particular types of interactivity.
   • Aesthetics: colours / corporate image.
   • Usefulness: who is it for (employees or clients or both)? Useful features.
   • User-friendliness: how easy is it to navigate?

4 **Task:** The learners evaluate their company web-site.

5 **Post-task:** The learners report back. Encourage the learners to consider their site as a resource to help their study of English e.g. by providing data for a company presentation, relevant vocabulary etc.

Variations

For the intro, issue the worksheet 'New media' or 'Your company and its web-site' (See Chapter 6.)

Acknowledgement

The authors would like to express thanks to Gavin Dudeney, particularly for the web-site evaluation worksheet (*The Internet and the Language Classroom*, 2000, p. 171, Cambridge University Press).

Company web-site evaluation sheet

Company: **Web-site address:**

Site description

Aesthetics

Usefulness

User-friendliness

Overall evaluation

11 Web-site familiarisation

Aim To exploit learners' company web-sites to practise fluency and vocabulary.

Procedure

1 **Intro:** Tell the learners that during their course they will make use of the World Wide Web to help them improve their Business English skills. The following activity will help them 'read' different types of web-site and help them focus on obtaining the information they need. Check that learners are familiar with the following terms used to talk about web-sites:

   - button
   - menu
   - navigation
   - link.

2 **Pre-task:** Ask the following questions:

   - What basic menu items would you expect to find on the homepage of a company's corporate web-site?
   - What menu items and links would you expect to find on the homepage of a news web-site?

3 **Task:** Tell the learners to visit the following web-sites to see if the menus match their lists:

   http://www.unilever.com.html

   http://www.daimlerchrysler.de/index_e.htm

   Then ask the learners to compare their lists with two of the most popular news web-sites in the world:

   http://news.bbc.co.uk

   http://www.cnn.com

4 **Post-task:** The learners report back.

12 Dictionary comparisons

Aim

To allow learners to explore the features and benefits of English – English dictionaries in different formats viz. paper-based / CD-ROM and web-based.

Level

Intermediate and above

Time

45 minutes

Procedure

1 **Intro:** Ask the learners which dictionary they use now. What are the good points? What do they see as the advantages of using a mono-lingual dictionary?

2 **Pre-task (optional):** Use a beamer to present features of electronic dictionaries.

3 **Task:** Divide the learners into three groups. Give each group a worksheet, and ask them to complete the relevant section.

Note: Two possible worksheets are included. 'Dictionary comparison sheet (1)' is for the full task. 'Dictionary comparison sheet (2)' allows learners to only compare the book, CD-ROM and the web-site.

[1]By permission of the publisher. From Merriam-Webster OnLine, www.Merriam-Webster.com, by Merriam-Webster Incorporated.

- **Group one:** Two paper-based dictionaries e.g. *Oxford Advanced Learners' Dictionary* (OUP); *Longman Contemporary English* (Longman)

- **Group two:** Two CD-ROM dictionaries e.g. *Cambridge Learners Dictionary on CD-ROM* (CUP); *Macmillan English Dictionary (MED)* (Macmillan)

- **Group three:** Two web-based dictionaries (e.g. *Cambridge* http://dictionary.cambridge.org/ and *Merriam Webster* http://www.m-w.com/home.htm[1]).

4 **Post-task:** The learners give feedback on features they have discovered and discuss the relative merits of each type of dictionary.

Comment

Depending on the learners' level of technical literacy, the teacher may need to monitor during the task phase, or suggest specific features to look for.

The 'Filters' are a superb feature of electronic dictionaries, but are not particularly user-friendly. The teacher may wish to suggest that learners explore this feature in a post-lesson task.

Rationale

Certain features of electronic dictionaries and web-dictionaries (e.g. search) are useful for learners. Awareness of these benefits can enhance autonomous learning.

Dictionary comparison sheet (1)

Consider the features / ease of use etc.

1. Paper-based

Dictionary:	Dictionary:
Comments:	Comments:

2. CD-ROM

Dictionary:	Dictionary:
Comments:	Comments:

3. Web-based

Dictionary:	Dictionary:
Comments:	Comments:

Dictionary comparison sheet (2)

Consider the features / ease of use etc.

1. Paper-based

Dictionary:

Comments:

2. CD-ROM

Dictionary:

Comments:

3. Web-based

Dictionary:

Comments:

13 Learner modalities

Level	Intermediate and above
Aim	To raise awareness of learner modalities (which sense learners favour for learning and remembering) and to reflect on some of the benefits of using a web-based, multimedia environment.
Time	60 minutes
Procedure	1 **Intro:** Write 'VAK' on the whiteboard and ask learners to guess what it means. (By giving them the first letter, 'V stands for Visual', learners can often guess the others). Write up: Auditory / Kinesthetic. Elicit some descriptions for each: e.g. visual learners like visual clues / reading instructions. Auditory learners like the teacher to read instructions. Kinesthetic learners 'learn by doing'. Ask which way the learners themselves take in information? How do they know? (e.g. they like the teacher to write things down).

2 **Pre-computer:** Tell the learners that they will now evaluate some web-based teaching materials. Tell them to explore the materials at will, and be ready to explain which aspects of the materials they enjoyed using. They should be ready to describe any choices they made.

3 **Task:** The learners explore a multimedia site on the Internet. E.g. the demonstration of web-based material on http://www.linguarama.com or other Business English materials such as http://www.globalenglish.com or http://www.york_associates.co.uk.intro.htm

4 **Post-computer:** The learners present what they liked and disliked about using the materials. Ask the learners if they had any insights into how they are as a learner?

Comment	The teacher can suggest some beneficial aspects about learning from multi-media e.g. choice of how many times to listen to an audio clip; listening and reading together; doing a mix and match activity on-screen etc.

14 Continuing to Learn

Aim

To evaluate some web-sites which may be useful for learners to continue with their language studies after the language course.

Procedure

1 **Intro:** Ask the learners what they are intending to do after the course has finished. Brainstorm ways of improving various aspects of the language.

2 **Computer task:** Ask the learners to evaluate some sites by giving them a 'star rating' for usefulness in post-course work. The teacher can issue the 'Universal worksheet' (see Chapter 6.). Possible sites to explore are:
Virtual business worlds http://www.bized.ac.uk/virtual/
Translation http://www.babel.altavista.com
Grammar http://www.englishclub.net

3 **Post computer:** The learners report back. The teacher inputs further ideas, depending on the specific needs of the group. E.g. for listening, try the BBC News or CNN audio; for writing, join a forum and post messages to other users in English. Issue the handout 'Continuing to Learn'.

Comment

This activity could form part of a wider session on ways of continuing how to learn – looking at language skills, business skills etc.

Continuing to Learn

There are many ways of using the Internet to help continue your contact with the English language after your course.

Grammar	Do some grammar exercises on the Web http://www.englishclub.net Use a grammar clinic http://www.lydbury.co.uk/discus/index.html
Vocabulary	Use an on-line English dictionary. For a list of specialist dictionaries, try: http://www.lib.uwaterloo.ca/libguides/1-10.html http://www.sciencekomm.at/advice/dict.html
Listening	Listen to the BBC News http://news.bbc.co.uk Listen to CNN audio http://www.cnn.com/audio
Writing	Join a forum and post messages in English to other users. Try: http://messages.yahoo.com/index.html
Reading	Read a web-page with software such as the *Oxford Genie* (OUP) or the 'QuickSearch' facility on the *Macmillan English Dictionary*. Search for a text in your field using a search engine such as *Google* http://www.google.com
Phonology	Listen to the phonemic symbols in English http://www2.arts.gla.ac.uk/IPA/ipa.html
Business skills: Presentations	Look up information about giving effective presentations http://www.research.ucla.edu/era/present
Business skills: Negotiations	Look up information about negotiating htpp://www.negotiationskills.com http://www.johnmole.com
Business English examinations	Consider taking a Business English exam http://www.eflweb.com/

6 Lesson activities

Chapter 6 is divided into the following areas:
1 Teaching with the Web
2 Teaching activities

1 ▦ Teaching with the Web

1.1 Why use the Web?

This chapter includes a number of practical teaching ideas. These ideas are often based on core principles such as 'search to research', which teachers can use when exploiting the Web in language lessons. Teachers should adapt the ideas to their individual teaching circumstances.

Search to research | Both learners and teachers can look for suitable authentic texts which they wish to discuss. Teachers can use the Web to build up company profiles of their key clients. This could range from researching for a full-blown project to searching for a piece of jargon which needs clarifying or viewing in context.

Spontaneous tasks | The learner may mention something, and it occurs to the teacher that they can access this information on the Internet, either there and then or at a future time.

Real world tasks | The Web can be used during a course to enable learners to perform tasks which they would do in real life. Learners in the UK needing to travel at the weekend can use http://www.railtrack.co.uk to plan out the route. Learners about to embark on a business trip but do not have a hotel can use the Web to explore possible places to stay.

Translation tools | The Web includes a vast number of sites which have been translated. Some corporate sites give you the option to click on an English version as well as an original language version. If the learner asks: 'What is this in (German)?' the teacher with little or no German can suggest he / she visit the company site to check the answer and report back.

Texts | Discussion lessons can be based on suitable text printed off from the Web. The learner can be involved in the selection of the text. (See: Activity 24, *Collocation search*.)

Uncovering grammar | It is possible to use the Web as a vast repository of text as a way of raising learners awareness of language patterns, the frequency of collocates etc. A specific example of using the Web to 'uncover grammar' (a term used by Scott Thornbury in *Uncovering Grammar*, 2002, Macmillan Heinemann) can be found in

Activity 15, *Grammar search* (modals).

Use *Google* at: http://www.google.com and enter the words 'English grammar' followed by the name of the area of grammar you wish to research. The number of 'hits' received will be an indication of the frequency of occurrence of that particular stretch of language.

1.2 Internet competencies

This section provides a handy checklist of useful technological competencies to have as a teacher wishing to integrate the Internet into a language lesson. It is roughly graded to follow an increase in skills from beginner to advanced.

Basic skills

- getting connected to the Internet and identifying when you are connected
- knowing the password which allows access to the network (if applicable)
- knowing which browser and e-mail software is installed on your computer and how to open them
- identifying the printer to which your computer is connected and how to switch it on and load paper.

Web browsing skills

- identifying hyperlinks (text, images)
- identifying and using various types of menus on web-pages
- using basic browser functions, i.e. back, forward, print etc.
- accessing and creating favourites or bookmarks
- finding things on the Web using search engines and directories
- using multiple browser windows
- using the browser history function to surf off-line
- using plug-ins such as a media player and *Adobe Acrobat Reader*®.

Other Internet skills

- composing, sending and receiving e-mail
- using other Internet communication software such as *NetMeeting*
- troubleshooting problems.

Using the Internet in language teaching

- using material from web-sites in a language lesson
- using a web-site and e-mail in a language lesson
- being prepared for technical problems
- using Web audio and video
- saving web-pages for later use.

Creating teaching materials

- copying and pasting text and images from web-sites into word processor documents

- producing customised worksheets using credited Internet derived materials
- checking that the web-sites proposed are current and live.

Advanced Internet skills
- creating web-pages using office software, e.g. Microsoft *Word*
- creating web-pages and web-sites using dedicated Web design software.
- creating interactive web-pages
- uploading web-sites to the Web
- running an on-line lesson with remote participants.

1.3 Using the Web: a checklist

There are a number of things to consider before teaching a language lesson which involves the use of the Web in some way. This checklist provides a useful list of points to consider.

Teacher's Internet checklist	
Do I have a clear activity in mind?	☐
Does the learner understand the activity and its purpose?	☐
Have I selected the web-sites I want to use?	☐
Do I know the addresses of these web-sites?	☐
Have I visited them beforehand to see whether they are 'live'?	☐
Does the computer I intend to use have the necessary hardware, e.g. speakers, and plug-ins, e.g. a media player?	☐
Does the printer work and have sufficient paper?	☐
Do I have a back up lesson plan in case the web-site I want to access will not load or there is some sort of equipment failure?	☐

1.4 Challenges and solutions

There are many challenges facing teachers wishing to incorporate the Internet into their teaching programmes. Here are some, along with some possible solutions.

Figure 6.1
Challenges and
solutions

Technical unreliability

Challenge	Solutions
You are unable to establish a link to the Internet. For example the broadband connection may be down.	Have a back-up lesson plan. Sometimes just having print-outs of certain web-pages allows you to proceed. Look for the sites before the lesson. This stores the pages visited in the computer's cache memory. Have some machines available 'in reserve'. This is possible when there are more computers than learners.

The web-site is not available

Challenge	Solutions
A web-site may be unavailable because the site is dead or due to problems with a remote server. Sometimes the web-site has been re-designed or changed in some way which makes the lesson redundant.	Check the sites you are going to use just before the lesson. Consider printing off a page from the desired site, which can be used as a back-up. Use the search engine cache.

Learner attitude

Challenge	Solutions
The learners are machine phobic, under-confident or simply unwilling to sit in front of a computer during what they perceive as a face-to-face language course. Some learners use the Internet all day and therefore do not wish to continue using the Internet during their studies.	Ensure that the activity is thoroughly integrated into the language lesson; has clear aims and outcomes. Make the aims explicit. Listen to the learners' preferences. If they prefer not to use a computer, use it sparingly, if at all.

Learner expectations

Challenge	Solutions
Learners may be over-awed by the volume of vocabulary they are exposed to. Or they may expect to understand and learn every word or phrase.	Separate the task from the material – consider real-world tasks: skimming and scanning, summarising etc. Encourage the use of an English – English dictionary in self-study time.

For more specific technical problems and advice see Chapter 9.

2 ▨▨▨▨▨▨▨ **Teaching activities**

The following bank of practical ideas is designed to provide off-the-shelf activities. We offer these ideas largely as templates which should be adapted to individual teaching situations, such as class-size, availability of computers, learner attitude and background, business area and so on.

Activities

Language areas

Speaking
1 New media
2 Virtual city tour
3 E-commerce
4 Travel plans
5 Bargain hunting

Listening
6 – 9 BBC world news

Reading
10 Current projects

Writing
11 Your company and its web-site

Integrated skills
12 Universal worksheet

Grammar
13 Company description (Present simple)
14 Share prices (Present continuous)
15 Grammar search (Modals)
16 Product comparison (Comparatives)
17 Airline (Present simple)
18 Time-line (Past simple)
19 Changes (Present perfect)
20 Predicting the future (*will*)
21 Consultancy (Conditionals)
22 Factory tour (Passives)
23 Missing articles (Articles)

Vocabulary
24 Collocation search

Business areas

Presentations
25 Presentations
26 The new product

Key

Each activity is laid out in the same way as explained below.

This symbol explains which language or business area is particularly focussed on in the activity.

This symbol is included if learners need to go on-line *during* the activity.

Aim	This states the activity's aim.
Level	All levels are approximate. Most of the ideas are multi-level, and the tasks can generally be adjusted to suit higher or lower levels.
Time	The time is an approximation, depending largely on learner reaction and interest. Discussion lessons, in particular, can take much longer.
Rationale	This is sometimes included to draw attention to a particular feature of the Web or a pedagogical point which has led to the creation of the task.
Procedure	1 **Pre-lesson task:** This draws the teacher's attention to something which should be done before the lesson – typically checking if a specific site exists, or printing off material before the lesson.
	2 **Intro:** This offers a warm-up to the topic in general.
	3 **Pre-task:** This involves setting the learners up with a specific task.
	4 **Task:** This describes the central part of the lesson.
	5 **Post-task:** This involves the learners re-grouping and reporting back to the teacher. It *always* assumes that the teacher will give language feedback after the report back, perhaps using an in-house feedback sheet, or a standard feedback sheet from a Business English resource book.
Extension	This outlines ways in which the activity could be developed.
Variations	This offers other activities which can be done instead, perhaps with different sets of learners.
Comments	This adds additional information of interest to the teacher, such as 'works well with learners who are all from the same company'.

1 New media

Language areas	SPEAKING LISTENING READING WRITING INTEGRATED SKILLS GRAMMAR VOCABULARY
Activity at a glance	**Learners complete a questionnaire about using the Internet, and have a discussion based on their responses.**
Aim	To develop fluency
Level	Intermediate and above
Time	45 minutes
Rationale	This activity is designed to get learners talking about the use of the Internet and how relevant / important it is.
Procedure	1. **Intro:** Ask the learners to tell you how many hours they spend on-line each week and write the results on the whiteboard. Tell the learners they will be discussing the use of the Internet.
	2. **Task:** Issue the 'New media' worksheet. The learners complete the sheet individually.
	3. **Post-task:** The learners can discuss their ideas in pairs first. They then report back to the whole group. The teacher monitors the resulting discussion.
Variation	The learners could do a S.W.O.T. analysis of the Internet.
Comment	A useful activity to do at the start of a Business English course to help gauge learner interest in the subject.

New media

Answer the following questions.

1. How often do you use the World Wide Web?

2. What do you generally use the Web for?

3. How often do you use e-mail?

4. Who do you use e-mail to communicate with?

5. What are the most useful / your favourite web-sites and why?

6. Have you ever purchased anything on-line? If not, why not?

7. Have you ever used a chat room or discussion forum? If yoo, what for?

8. What is the most frustrating thing about the Internet?

9. What type of business is most likely to be successful on the Internet? Why?

10. Do you prefer to get your news from the Web or from 'old' media such as newspapers and television? Why (not)?

2 Virtual city tour

Language areas	SPEAKING · LISTENING · READING · WRITING · INTEGRATED SKILLS · GRAMMAR · VOCABULARY
Activity at a glance	**Learners use the Web as a research tool to prepare a short presentation on their home towns.**
Aims	To allow learners to present their home towns, using the Web as a resource and to encourage fluency and vocabulary development.
Level	Intermediate and above
Time	90 minutes
Rationale	Based on the principle that 'search to research' tasks are very motivating. The 'knowledge / experience gap' in a multi-lingual group is usually high, and learners generally wish to know about each others' countries and cultures.
Procedure	1. **Intro:** Ask the learners if they have ever visited each others cities. What kind of things would they like to know about each city? Collate a list of their replies on the whiteboard.
	2. **Pre-computer task:** Issue the worksheet 'Virtual city tour' and check the learners are comfortable with the vocabulary and the task.
	3. **Task:** At the computer, the learners complete each section of the worksheet, filling in the information for their own city, using the Web as a research tool. The teacher can mention useful travel sites, or check that the learners can use a search engine.
	4. **Post-task:** The learners deliver the presentations of their own cities.
Comment	Highly appropriate for Business English learners who need to arrange stays for visitors, organise itineraries etc.
Variations	Other presentation-style tasks, using the Web as a resource. Encourage the learners to cut and paste text and graphics onto OHTs, or create a handout.

Vitual city tour	
City:	
Data	**Famous buildings / monuments**
Size (area)	
Population	
Location	
Climate	
History – key dates	**Food – specialities / restaurants / drinks**
Culture – festivals / customs	**Other key information e.g. major industries**

3 E-commerce

Language areas	SPEAKING · LISTENING · READING · WRITING · INTEGRATED SKILLS · GRAMMAR · VOCABULARY
Activity at a glance	**Learners discuss the area of e-commerce. They explore the facilities offered by an on-line trader such as Tesco.**
Aims	To develop fluency
Level	Intermediate and above
Time	45 – 60 minutes

Procedure

1. **Intro:** Brainstorm words in English which start with 'e-' like *e-mail*. Examples: *e-business / e-commerce / e-book / e-learning*.

2. **Pre-task:** This lesson will look at the way the Business English world has changed with the growth of e-commerce. The teacher reveals the following questions, using a flip-chart:
   • Does your company sell on-line?
   • What are the benefits of selling on-line?
   • Do you see any dangers?

3. **Task:** The learners visit the following site: http://www.tesco.co.uk. They prepare a short report describing the on-line buying part of the site and details of the payment method.

4. **Post-task:** The teacher leads the follow-up discussion. Discussion points could include:
   • Has your view on e-commerce changed having looked at the sites?
   • Predict the future of e-commerce.

Extension

Compare the service offered by Tesco with similar providers in the learners' own countries.

Variations

Alternative intro questions:
• Have you ever paid for anything on-line?
• If so, did the transaction proceed smoothly, or did you have any problems?

Learners look at an on-line bank (e.g. http://www.egg.co.uk) and the teacher asks:
• Why has there been a rise in on-line banking in the UK?
• Has there been a similar rise in your country?
• How do you see the future of on-line banking?

4 Travel plans

Language areas	SPEAKING · LISTENING · READING · WRITING · INTEGRATED SKILLS · GRAMMAR · VOCABULARY
Activity at a glance	**Learners plan a trip by using travel information web-sites.**
Aims	To discuss travel plans, using travel and government web-sites.
Level	Intermediate
Time	60 minutes
Web-sites	Travel Guides: http://www.lonelyplanet.com and www.roughguides.com
	Official Information: http:///www.fco.gov.uk/travel and http://www.who.int/ith

Procedure

1. **Intro:** Tell the learners that they are colleagues attending an international conference in a country other than their home country. After the conference they are planning to spend three extra days together being tourists in that country. Brainstorm the type of information the learners are going to look for, especially if you have more than one group.

2. **Pre-computer task:** Agree with the learners about how they are going to organise their planning. Do they intend to look at the information on the web-sites and see what comes up or are they going to decide what they need to know and go looking for it? Guide the learners towards more exotic locations. They should choose a country that none of them has visited before. An atlas or a map of the world might be useful.

3. **Task:** The learners collect the agreed information about the chosen country/countries. They create their rough schedule and consider any potential problems and their solutions.

4. **Post-task:** The learners reconvene and present the results of their research and discussion. If there is more than one group, each group can question the other about the feasibility of their plans.

Comment

It is important to clarify the scope of the activity from the outset.

Variations

This activity can be a longer project, spanning a day or parts of several days. In this case, the schedule would progress beyond being rough and the learners could investigate hotels, restaurants, tour companies etc. This could culminate in a group presentation.

Travel plans

You and a group of colleagues are attending an international conference in

.... After the conference you plan to spend three

extra days together being tourists in that country.

Use the web-sites suggested by your teacher to find information and advice about where to visit in the country and where to avoid; what to pack; what to buy; what to do in the case of emergencies.

Prepare a rough schedule for the three days which includes contingency plans for any possible problems.

5 Bargain hunting

Language areas	**SPEAKING** · LISTENING · READING · WRITING · INTEGRATED SKILLS · **GRAMMAR** · **VOCABULARY**
	COMPARATIVES

Activity at a glance	**Learners use the Internet to look for good value purchases on the Web.**
Aims	To encourage fluency and vocabulary development; to practise numbers and adjectives of comparison.
Level	Pre-intermediate and above
Time	45 minutes
Procedure	1. **Intro:** Ask the learners how their company selects its suppliers. Is price the most important consideration? What factors other than price do they take into account?
	2. **Pre-computer task:** Issue a newspaper. Ask the learners to look at the prices of 4 – 5 selected electronic articles in a high street store advertisement. Note down these prices.
	3. **Task:** The learners go to a site such as http://www.empiredirect.co.uk Here, they check the prices of the items selected.
	4. **Post-task:** The learners present their findings, comparing the figures. They discuss whether they would buy the articles on the Web. Why? Why not?
Extension	The learners could also do a web search, in order to find out if there are any cheaper suppliers.
Variations	**Role play:** your boss has asked you to buy a new mobile phone on the Internet. Do a search and come back to him / her with recommendations on three common models. List their prices and features.
	Insurance search: Look at two holiday insurance companies and compare their quotes.

6–9 BBC world news

Language areas	SPEAKING **LISTENING** READING WRITING INTEGRATED SKILLS GRAMMAR VOCABULARY
Activity at a glance	**Learners use either video or audio clips on the BBC site to work on listening. There is a choice of four worksheets focussing on different listening skills (summarising, deepening, updating, widening).**
Aim	To promote authentic listening based on current, international (and therefore, hopefully, familiar) news stories.
Level	Intermediate – advanced
Time	10 – 60 minutes depending on the combination of activities used and the level of the learners.
Rationale	Learners can be in charge of the controls and pause and repeat certain sections at will. Learners receive exposure to normal speaking speed, with extra pauses inserted at will.
Web sites	http://news.bbc.co.uk/, the BBC news frontpage and http://www.bbc.co.uk/worldservice/news/summary.ram, a direct hyperlink to the BBC World Service Radio world news summary. This is a hyperlink to a *RealPlayer* audio file and so will automatically open the *RealPlayer* plug-in, provided it is installed on your computers.
Procedure	1. **Pre-lesson:** For the worksheets 'BBC world news: updating' and 'BBC world news: widening', check the hyperlinks to previous reports and that outside web-sites are live.
	2. **Intro:** Ask about the learners' experience of getting news from the Web. Have they used any web-sites? Do they prefer the Web to 'old' media such as newspaper / radio / TV? What do they like / not like about the news delivered by the Web?
	3. **Pre-computer task:** Explain that they are going to use the BBC's news web-site for a listening activity. Check if anyone is familiar with this web-site. If so, ask them to give a brief description. Otherwise, elicit from the learners a list of the items they would expect to see on a news web-site menu. Show the learners the frontpage so they can see the main menu on the right of the web-page. Elicit any of the stories that the learners might expect to hear on that particular day.
	4. **Task:** Introduce the learners to the equipment and the software. If someone is familiar with *RealPlayer*, ask them to demonstrate how to use it. Otherwise, do the demonstration yourself. Issue the appropriate worksheet depending upon the skill to be worked on. **Summarising** Tell the learners to listen to the 5 minute news summary and take notes on the headlines; any details of the story and, perhaps, vocabulary to look up or ask the teacher.

Deepening Tell the learners to chose two of the stories and use the BBC news web-site to locate the written equivalent of these stories. Read them, take notes and compare with the audio version. Listen again after reading.

Updating Tell the learners to look to the right of the written version of the story. If the story is on-going the BBC will list hyperlinks to previous reports on the issue. Go to these reports and construct a chronology of how events arrived at the current situation.

Widening Also on the right of the written version of the story, the BBC sometimes provide hyperlinks to the web-sites of people and organisations connected to the story. Tell the learners to visit these web-sites and find information to supplement, support or contradict the BBC's report.

5. **Post-task:** The learners reconvene and verbalise the results of their research. These can then be worked into presentations or written reports. (With pictures and other information such as charts or tables of figures, copied and pasted onto transparencies, *PowerPoint* slides or pages.)

Extension

If you have a CD-ROM or other type of electronic dictionary installed on the computers being used, e.g. *Cambridge Learner's Dictionary* or *Oxford Advanced Learner's Dictionary*, remind the learners that they can use this simultaneously to check vocabulary.

Comment

There are a number of activities that can be done using this web-site. The worksheets can be done individually, in the suggested sequence, over a period of time or divided between members of a group.

Variations

If you have learners who prefer bite-sized pieces of authentic listening and who like to see transcripts, there are other services available from the BBC.

The BBC's World Service web-site takes a single international news story every Monday, Wednesday and Friday. This is presented with a brief summary; the audio clip and its transcript and a list of vocabulary from the transcript along with an audio clip giving pronunciation models. There is an archive of these exercises going back to 1999. See: http://www.bbc.co.uk/worldservice/learningenglish/news/index.shtml

This same web-site also has a section on business vocabulary which is supported by business radio programmes, again with audio and transcripts. See: http://www.bbc.co.uk/worldservice/learningenglish/work/index.shtml

Both of these web-pages, as well as the BBC news exercises can be recommended as continuing to learn activities. (A computer with a 56k modem and a recent version of *RealPlayer* will have few problems downloading and playing the audio clips provided by the BBC's web-site.)

Comment

It may be necessary to introduce the learners to *RealPlayer*, or check if any of the learners can demonstrate how to use it.

BBC world news: summarising

Listen to the World Service news summary from the BBC's news web-site. Your teacher will give you the relevant addresses.

Use the table below to make notes on today's stories. (There may not be as many as eight.) Use the first column to record the main subject of each story; the second column to give a summary of the main points and the last column to record any new vocabulary or any words / phrases you wish to ask your teacher about.

Listen as many times as you want but remember that the target is not to understand every single word but rather to be able to summarise the content.

Headline	Main details of story	Vocabulary

BBC world news: deepening

Go to the BBC's news web-site. Choose one or two stories from the news summary.

Find the written report relating to the stories you have chosen. Compare this with your notes from the summarising activity.

Story 1

What I heard	What the report says

Vocabulary

Story 2

What I heard	What the report says

Vocabulary

BBC world news: updating

Go to the BBC's news web-site and choose one of the stories from the news summary. Find the written report relating to this story and use the links on the BBC's news web-site to trace its history.

See how the story has developed and summarise as much as possible.

Date	Summary of events
Today	

BBC world news: widening

Go to the BBC's news web-site. Choose one of the stories from the news summary and find the written report relating to this story. Use the links on the BBC's news web-site to investigate other web-sites connected with your chosen story.

Did you find any additional information that was not included in the BBC's reports? Did you discover a different point of view to the one put forward by the BBC? Are the events of the story any clearer as a result of investigating these other web-sites?

Link	Additional information

10 Current projects

Language areas	SPEAKING	LISTENING	**READING**	WRITING	INTEGRATED SKILLS	**GRAMMAR**	VOCABULARY

PRESENT CONTINUOUS

Activity at a glance — **Learners describe current projects, using information from their companies' web-sites. This may be part of a language lesson which looks at uses of the present continuous.**

Aim — To give learners a reading task, based on material from their company web-site, and to provide free practice in the use of the present continuous for things happening at the moment.

Level — Intermediate

Time — 30 minutes

Procedure

1. **Intro:** Ask the learners about their company web-site. Does it carry news about current projects that the company is involved in?

2. **Pre-computer task:** Issue the worksheet 'Current projects' and set up the task.

3. **Task:** The learners select up to four current projects or events described by the web-site and what is happening at the moment. Print out anything which is of interest, for reading off-screen.

4. **Post-task:** The learners reconvene and present the results of their research. They can also indicate if they are involved in the project or affected by it in any way.

Extension — The learners do a follow-up task, going to a business news web-site to identify events in the world that are having an effect on the company and how the company is reacting. For example, conflict in the Middle East having an effect on oil prices which then impacts on oil companies, power generators, car manufacturers etc.

Variation — The learners could select a current project in which they are involved and then visit their company web-site to find more information to support a short presentation.

Comment — This may be part of a language lesson which looks at uses of the present continuous. It is important to check that the learners are familiar with the operation of the web-sites chosen and that they do provide information about current projects.

Current projects

Visit your company's web-site. Find up to four projects or events that are happening in the company at the moment. Fill out the table below.

Project / Event	What's happening at the moment?	Are you involved in this project / event and how?

11 Your company and its web-site

OPTIONAL

Language areas	**SPEAKING** LISTENING READING **WRITING** INTEGRATED SKILLS GRAMMAR VOCABULARY
Activity at a glance	**Learners complete a questionnaire about using their company web-site, which is used as a basis for discussion.**
Aim	To provide writing practice and to encourage fluency by discussing the Internet.
Level	Intermediate and above
Time	45 minutes
Rationale	This activity is designed to get learners talking about the use of the Internet and how relevant / important it is.
Procedure	1. **Intro:** Ask the learners to tell you if they have visited their company web-site, and if so, what they think about it.
	2. **Task:** Issue the worksheet 'Your company and its web-site'. If necessary, the learners visit their company web-site. They complete the questionnaire.
	3. **Post-task:** The learners can discuss their ideas in pairs first. They then report back to the whole group. The teacher monitors the resulting discussion.
Comment	A useful activity to do at the start of a Business English course to help gauge learner interest in the subject.

Your company and its web-site

Company: Web-site address:

1. How often do you visit your company web-site?

2. When was the last time you visited the web-site?

3. What do you use the web-site for?

4. Is the design of the web-site in line with the image of the company? Give details.

5. Is it well organised / easy to use? Does it contain useful / up-to-date / accurate information about the company and its products / services? Give details.

6. How does it contribute to the way the company does business?

7. Would you recommend customers / clients / anyone else to look at your company's web-site? Why (not)?

8. What suggestion(s) would you give to the designers of the web-site to make it better?

12 Universal worksheet

Language areas	SPEAKING LISTENING READING WRITING INTEGRATED SKILLS GRAMMAR VOCABULARY
Activity at a glance	**Learners use the computer during a lesson to research an area, in order to feedback later.**
Aim	Variable, depending on the task.
Level	Elementary and above, depending on the task.
Time	Variable, depending on the task.
Rationale	This sheet is designed to be used by teachers and learners for a variety of different tasks, usually of the 'search, find and report' variety.
Procedure	**Pre-lesson task:** Complete the form before the lesson / activity with the relevant web-site address.

Pre-computer task: Distribute the 'Universal worksheet'. Explain the purpose of the vocabulary box – to record any useful language that learners wish to retain after the activity, or to ask their teacher about.

Task: As chosen by the teacher (or learners).

Post-task: The learners report back on their findings. The teacher can follow-up on any useful vocabulary resulting from the task.

Universal worksheet

Lesson topic:

1. Look at the following site:

2. Task

3. Task result

Make notes here.

4. Vocabulary

13 Company description

| Language areas | SPEAKING | LISTENING | READING | WRITING | INTEGRATED SKILLS | **GRAMMAR** | VOCABULARY |

PRESENT SIMPLE

Activity at a glance
Learners use the Web to research factual data about a company, in order to give a simple company presentation.

Aim
To provide practice in the use of the present simple for statements and descriptions; to provide practice in the form of the present simple e.g. the third person 's'.

Level
Pre-intermediate

Time
45 minutes

Rationale
Many learners tend to over-use the present continuous, and this activity helps re-enforce the common grammatical pattern of the present simple.

This activity could form part of a longer lesson on the present simple, coming after a focus on form.

Procedure

1. **Pre-lesson task:** Print out the homepage from a company web-site which will be of interest to the learners. Check that the page contains simple statements.

2. **Intro:** Issue the worksheet 'Company description'. Draw the learners' attention to the occurrences of the present simple on the selected homepage. Ask them why it is used (e.g. expresses simple facts). Mention that many sites present a description of goods or services, and make claims for the company.

3. **Pre-computer task:** Write a number of words on the whiteboard which are sometimes used on a homepage to give a company description. Some possibilities are:

produces	offers	sells	promotes
exports	provides	supports	uses

Tell the learners they are to prepare a short presentation on a company of their choice. They should research a company that interests them. (It could be their own company, if appropriate.) Use the presentation template provided.

4. **Task:** The learners use the Web to find the information required for their presentations.

5. **Post-task:** The learners give their presentations. The teacher gives feedback, focusing on the present simple.

Acknowledgement
The authors would like to express thanks to Mark Powell, particularly for 'A ten-point presentation plan' (*In company*, (2000) p. 53, Macmillan).

Company description

Research a company on the Web and use this worksheet to prepare a short presentation.

Introduction	Good morning ladies and gentlemen. I'd like to talk to you about ...
The company (has ... branches / factories) (headquarters are ...)	
The product / service (sells / produces / makes ...)	
The work-force (MD is ...) (employs ...)	
Other information	
Finish	Thank you for your attention. Do you have any questions?

14 Share prices

Language areas	SPEAKING	LISTENING	READING	WRITING	INTEGRATED SKILLS	**GRAMMAR**	VOCABULARY

PRESENT CONTINUOUS

Activity at a glance	**Learners practise the present continuous by describing what is happening to share prices at that moment, using the web-site of a major stock market.**
Aim	To practise using the present continuous for current and changing situations, using web-sites which give stock market information.
Level	Intermediate and above
Time	30 minutes
Rationale	This task is particularly useful for learners who work in finance.
Web-sites	London Stock Exchange http://www.londonstockexchange.com/
	New York Stock Exchange http://www.nyse.com
	Nikkei http://www.nni.nikkei.co.jp/
	Bloomberg http://www.bloomberg.com

Procedure

1. **Intro:** Ask the learners if they invest in shares. If so, do they keep track of their investments via the Web? Ask if anyone knows what the current trend of the stock markets is and why.

2. **Pre-computer task:** Familiarise the learners with the layout of the web-sites you intend to use and where the relevant links are located.

3. **Task:** The learners, individually or in pairs, access the web-sites listed above and collect information on what the market is doing at the moment.

4. **Post-task:** The learners report back to the group and present and compare their findings.

Extension	If the reason for the general trend was not established or is unclear, it can be discussed. The learners offer their opinions supported by information from the web-site they used.
Variation	The same activity can be carried out with any commodity whose price fluctuates from hour to hour, e.g. oil prices, especially if this is relevant to the learners' jobs.
	This activity can also be used during a lesson on the language of trends.
Comments	If the learners are experienced in using this type of web-site they can also find out what the shares of significant companies are doing, especially those of their own company, or of personal portfolios if they are quoted on the stock market in question.
	This task has more real world relevance if you select stock markets that are open at the time during which the lesson is taking place, i.e. if you are in Europe, the New York Stock Exchange does not open until the afternoon, so London may be a better option.

15 Grammar search

Language areas	SPEAKING · LISTENING · READING · WRITING · INTEGRATED SKILLS · **GRAMMAR** · VOCABULARY
	MODALS

Activity at a glance	**Learners use a search engine on the Web in order to explore meanings of modal verbs.**
Aim	To investigate how various modal verbs are used by seeing them in context.
Level	Intermediate – advanced.
Time	Up to 60 minutes
Rationale	Allows learners to see a stretch of language in an authentic setting, in order to check their grammatical assumptions.

Procedure

1. **Intro:** Brainstorm modal verbs. Check the learners' understanding of how the various modal verbs (of obligation / possibility) are used and in what contexts. Welcome any disagreement or ambiguity about usage and pose them as questions to be answered by the activity.

2. **Pre-computer task:** Find out whether the learners are familiar with using quotation marks to conduct a search with *Google* for an exact phrase. Give a short demonstration. Select a simple noun-verb combination into which various modal verbs can be inserted. For example, "euro will rise"; "euro might rise"; "euro may rise" etc. Allocate a modal verb to each learner to investigate.

3. **Task:** The learners use http://www.google.com to search for the agreed phrases. They analyse the results to determine the context in which the chosen modal verb has been used. Sometimes this can be done from the *Google* result; sometimes the web-page found by the search has to accessed in order to clarify the context.

4. **Post-computer:** The learners report back to the group and decide whether the differences of opinion which emerged before the search have been resolved and the questions posed answered.

Variation	Clearly this activity can be performed using any area of language in which 'real-world' examples can help to clarify usage or apparent inconsistency in usage.
Comment	We recommend trying the searches yourself to determine whether your learners will find enough examples to work with.
	Note that since *Google*'s web crawlers can take several days to up-date the entire *Google* index you should not rely on that day's news stories appearing in the search results. Keep the subject of your search fairly general.
Acknowledgement	The authors would like to express thanks to Scott Windeatt, David Hardisty and David Eastment, for their idea '1.1 It all depends' (*The Internet*, 2000, OUP) which uses the Web in a similar way.

16 Product comparison

Language areas	SPEAKING LISTENING READING WRITING INTEGRATED SKILLS **GRAMMAR** VOCABULARY
	COMPARATIVES

Activity at a glance	**Learners use the Web to research details of their competitors' products or services. They then practise the language of comparisons.**
Aim	To practise the language of comparisons
Level	Elementary and above
Time	60 minutes
Procedure	1. **Pre-lesson task:** Check that suitable sites exist i.e. the learners' company and a competitor. It may be necessary to ask the learners for the names of their competitors before the lesson. Have the web-site addresses to hand.
	2. **Intro:** Draw a number of ovals on the whiteboard. Around each oval, brainstorm some common adjectives which your learners can use to describe their products. Some possibilities are:

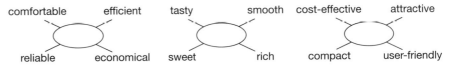

Ask the learners to briefly describe their products.

3. **Pre-computer task:** Elicit the name of a competitor for each of the learners' companies. Issue a 'Product comparison' sheet, and ask the learners to fill in the name of their company and a competitor, plus the site addresses if you have left these sections blank. The learners then complete the left-hand column with information about their product.

4. **Task:** The learners visit their company's competitor's site and complete the sheet.

5. **Post-task:** The learners give mini-presentations, comparing their own product / service with the competitor's.

Variation	The learners compare their actual web-site with a rival's site. More advanced learners can look at the persuasive element of the web-site, or the underlying corporate philosophy. Learners working in a group from the same company can each research a different competitor.
Comment	Select appropriate adjectives depending on the learners' products.

Product comparison

Your company:	Competitor:
Web-site address:	Web-site address:
Company details *(size, employees, turnover etc)*	**Company details** *(size, employees, turnover etc)*
Product / Service	**Product / Service**

17 Airline

Language areas	SPEAKING · LISTENING · READING · WRITING · INTEGRATED SKILLS · **GRAMMAR** · VOCABULARY
	PRESENT SIMPLE

Activity at a glance **Learners practise talking about schedules and timetables by comparing air ticket prices researched on airline web-sites.**

Aim To raise awareness of or provide free practice in the use of the present simple to talk about schedules.

Level Elementary – intermediate

Time 60 minutes

Rationale Many business travellers use an on-line service to check, compare and book airline tickets.

Procedure

1. **Intro:** Ask the learners which is their favourite airline and why.

2. **Pre-computer task:** Tell the learners that their company wants to save money by reducing travel costs. They have been given the task of researching the cheapest ways of reaching the destinations to which company personnel travel most often. Check that the learners are familiar with the operation of this type of web-site. Ask them to select which airlines they wish to compare. The starting point for the journeys should also be agreed as well as three destinations.

3. **Task:** The learners access the chosen airlines' web-sites either individually or in pairs and find the cost of flying from the agreed airport to each destination in both economy and business class. Once they have this information they can compare which airline is the cheaper and more expensive for each route and in each class.

4. **Post-computer task:** The learners regroup and present the results of their research. Based on the information they decide, collectively, which airline they should use on which routes. The teacher provides correction and, if necessary, some explanation of the grammar point.

Extension The learners might encounter special offers or other factors, such as group booking discounts, which might also have an effect on price. These offers can be quite complicated, and learners could try and explain how these work.

Comment It is better to choose a large airport from which most airline companies fly.

A list of airline web-sites can be found at:
http://uk.dir.yahoo.com/Business_and_Economy/Shopping_and_Services/Travel_and_Transportation/Airlines/

Airline

Your company is cutting costs – again! You have been asked to find out if it is possible to reduce the company's travel costs. Look at two airline web-sites and compare the ticket prices to various destinations as well as the difference between travelling economy and business class.

Airline:	Web-site:	
Destination	Economy class	Business class

Airline:	Web-site:	
Destination	Economy class	Business class

Comparison		
Destination	Economy class: your choice	Business class: your choice

18 Time-line

Language areas	SPEAKING LISTENING READING WRITING INTEGRATED SKILLS **GRAMMAR** VOCABULARY
	PAST SIMPLE (PASSIVE)

Activity at a glance	**Learners use the company history section of their web-site to research and give a short presentation.**
Aim	To review verbs in the past simple / past simple passive
Level	Intermediate and above
Time	30 minutes
Rationale	The site is usually interactive in some way and therefore it can be a motivating task for learners to look at the site. Many learners do not know about their company history, and may well be interested in obtaining information.

Procedure

1. **Pre-lesson task:** Visit the web-sites of the learners in your group to see whether it includes a company history. (If there is no company history, you may wish to adapt the lesson, considering one of the variations below.)

2. **Intro:** Write the following on the whiteboard:

founded – grew – expanded – merged – took over

   Ask the learners to supply key dates in the history of their company. If the learners do not know, this can form the 'information gap' used in the computer task.

3. **Task:** The learners look at their company's web-site on the Web. Using the time-line, they write down the key events.

4. **Post-task:** The learners give presentations. You may wish to focus on which phrases are past simple and which are passive.

Variations

If there is no company history on the site, you may wish to provide a model for the learners to then produce their own short company history. This could involve a web-search.

Learners can search for verbs in the past simple and use a highlighter. This is an excellent 'noticing' task, which is good for consolidation.

Prepare a set of cards based on the data researched on the learners' web-site. One set of cards has the year, the other has the corresponding event. Learners can then match them. This activity is worthwhile if it can be recycled with other groups from the same company.

Time-line: company history

Go to the web-site of your company and use the menu to find the company history. Add the main events to the time-line. Try to use as many of the verbs in the boxes as possible.

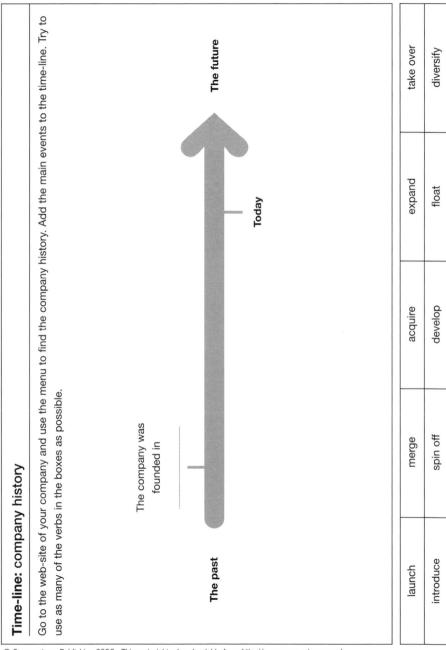

The past

The company was founded in
_____

Today

The future

launch	merge	acquire	expand	take over
introduce	spin off	develop	float	diversify

19 Changes

Language areas	SPEAKING LISTENING READING WRITING INTEGRATED SKILLS **GRAMMAR** VOCABULARY
	PRESENT PERFECT

Activity at a glance
Learners receive information on a topic, and then use the Web as a resource in order to update on recent events.

Aim
To review uses of the present perfect (e.g. *since* / *recently* / present perfect with current effect) and to contrast tenses (e.g. past simple vs present perfect).

Levels
Intermediate

Time
45 minutes (excluding watching authentic video)

Rationale
This activity capitalizes on the time-gap between say, a published article in a Business English course book, with a longish shelf-life, and the immediacy of information on the Web, and incoming reports on recent events.

Procedure

1. **Pre-lesson task:** Select an article, authentic video of interest, or text from a company brochure which provides a topic area of interest for the learners and where recent events have been in the news. E.g. Richard Branson in 'Clever Dick' from the series *Millionaires* (1998, Carlton UK Television).

2. **Intro:** Brainstorm how much the learners know about that particular topic e.g. Who is Branson and why is he famous?

3. **Pre-computer task:** Present the information about the company. E.g. learners watch the video or read the text and take notes, using the left hand column of the worksheet 'Changes'. The learners use their notes to give summaries to the teacher. Tell the learners that much has happened since the date when the video was made or the text was written, and that they will be using the Internet to explore what has happened.

4. **Task:** The learners perform a web-site search, looking for documents which give recent information, in order to gather information about what has happened since the video / text. They take notes in the right hand column.

5. **Post-task:** The learners give their presentations. The teacher gives feedback, particularly focusing on the use of tenses. Ask: 'What's happened since… / recently?'.

Variations
Articles from Business English course books, authentic videos or company brochures can be selected. Although the time element is exploited here, this activity can be adapted to explore different points of view on events.

Comments
This activity is particularly useful if there has been recent company news which supercedes an article in published Business English materials.

Acknowldegement
Based on ideas provided by Nina O'Driscoll, BESIG Conference, 1999.

Changes	
Past events	Update
In 1999 / in July / last year / last month	Since then / recently

20 Predicting the future

Language areas	SPEAKING · LISTENING · READING · WRITING · INTEGRATED SKILLS · **GRAMMAR** · VOCABULARY
	WILL

Activity at a glance	**Learners practise using *will* for making predictions by considering items on their companies' web-site news or press release pages.**
Aim	To provide practice in the use of *will* to make predictions about future events, using a relevant web-site.
Level	Intermediate
Time	60 minutes
Procedure	1. **Pre-lesson task:** Check that the learners' companies have suitable web-sites.
	2. **Intro:** Ask the learners about their company web-sites. Do they carry news about the companies and the market in which they operate?
	3. **Pre-computer task:** Issue the 'Predicting the future' worksheet and set up the task.
	4. **Task:** The learners select three current stories or events described by the web-site and summarise them. They then make a prediction about how they believe this situation will develop in the future. There is no limit to how far into the future they can predict.
	5. **Post-computer task:** The learners regroup and present the results of their research and their predictions. They can also explain why these particular stories attracted their attention and justify their predictions. The extent of these justifications will depend on level. The teacher provides correction and, if necessary, some explanation of the grammar point.
Extension	If the learners' predictions are over a short period, i.e. within the time period of the course, they can revisit the web-site and see how accurate they were. They can attempt to justify their predictions.
Variation	A quicker version involves reading only the headlines.

Predicting the future

Go to the following web-site: ....

Select three stories. Use the table below to summarise the current situation of each and your prediction for how these situations will develop in the future.

What's happening now	My prediction for the future
→	
→	
→	

Use some of these phrases to begin your predictions:

I think	I believe	In my opinion
I expect	I hope	I suppose
From my point of view	I predict	I wonder if

21 Consultancy

Language areas	SPEAKING	LISTENING	READING	WRITING	INTEGRATED SKILLS	**GRAMMAR**	VOCABULARY
						CONDITIONAL 2	

Activity at a glance	**Learners suggest options for spending a corporate language training budget.**
Aim	To allow practice in using the second form of the conditional
Level	Intermediate and above
Timing	60 minutes
Assumptions	Learners are able to construct second form conditionals. This activity could be used as a free stage of a grammar review lesson, or as a diagnostic stage in a Test-Teach-Test lesson.
Procedure	1. **Intro:** Write the following options on the whiteboard and ask the learners to decide which is preferable.

   • Take a two week language course in the UK.
   • Take a one year language course in-company, studying three hours per week.
   • Take a distance learning language course over the Internet.

   The learners discuss the various pros and cons.

2. **Pre-computer task:** Tell the learners that they work for a consultancy company. They have been asked to research a number of language schools on the Internet, and then advise their client (a large company based in e.g. Germany, wishing to train 100 employees) which school they should send their employees to. The learners should be prepared to justify their recommendation.

3. **Task:** This could be very open-ended or very closed.
   • An open-ended task could be: 'search the Web for language schools and take some details'.
   • A closed task could involve the teacher giving out three specific web-site addresses as follows: one for an international language school, one for a local language school and one for an on-line school.

4. **Post-task:** The learners give their presentation. The teacher gives language feedback on the target structure: *should / would*. The learners write a short consultancy report.

Alternatives	**Web-site design consultancy:** Brainstorm what makes a good web-site. Exchange ideas on any good sites the learners know. Tell the learners they have to make a short report on a number of sites which have asked for their site to be re-designed. The learners visit the sites and make suggestions and recommendations. They report back.

Investment consultancy: The learners adapt the 'Investing in the future' task and recommend companies to a client wishing to invest a specified sum of money.

Management consultancy: The learners take a company in trouble and research details on the Internet; they then make suggestions to the Board of Directors. The choice of company could be driven by current events and scandals!

22 Factory tour

Language areas	SPEAKING LISTENING READING WRITING INTEGRATED SKILLS GRAMMAR VOCABULARY
	PASSIVES

Activity at a glance	**Learners use the Coca Cola site to explore an interactive manufacturing process, in order prepare a similar factory tour based on their own product.**
Aim	To review the passives for processes, to develop fluency and to build vocabulary
Level	Intermediate and above
Time	30 minutes
Procedure	1. **Intro:** Issue cards with key phrases such as the example below. Tell the learners that this is an overview of the factory process involved in, say, how Coca Cola is made. The learners put the cards in the correct order. If any words are new, learners can guess their meaning, and check their guess from the site.

despatched	mixing and blending	capping
washing and rinsing	warehouse and delivery	coding
inspection	ingredient delivery	packaging

2. **Task:** The learners check their answer by visiting the 'Virtual plant tour' at Coca Cola's site: http://www.vpt.coca-cola.com/vpt_index.html

   Depending on the level of learner interest, they explore the interactive tour itself.

3. **Post-task:** The learners use an OHT to create their own virtual factory tour.

Variation	Use a beamer to present the process of making Coca Cola. The learners check the order in which they put the cards.
Extension	There are a number of plant tours on the Internet. These are listed at: http://bradley.bradley.edu/~rf/plantour.htm

The learners could explore a tour related to their area, as a prelude to preparing their own presentation. |
| Comments | This task is particularly for use with learners involved in the manufacturing of a product. Note that there are more stages in the process on the web-site than mentioned here. |

23 Missing articles

Language areas	SPEAKING · LISTENING · READING · WRITING · INTEGRATED SKILLS · **GRAMMAR** · VOCABULARY
	ARTICLES

Activity at a glance	**Learners consider the use of articles by re-constructing a text taken from the Internet which has had the articles removed.**
Aim	To raise awareness in the use of articles.
Level	Intermediate.
Time	15 minutes
Rationale	Useful for learners from language backgrounds who have difficulties in this area, such as Polish, Japanese and Arabic learners. Takes advantage of the electronic nature of text in that the teacher can quickly delete articles from the original.
Procedure	1. **Pre-lesson task:** Select a short paragraph from a suitable site. Cut and paste the article into a *Word* document. Delete the articles. Retain a copy of the original article.
	2. **Intro:** Remind the learners that successful use of articles (*a* / *an* / *the*) in English includes the concept of the 'absence of an article' as well as a sound understanding of countable and uncountable nouns. Review common rules / guidelines, if appropriate.
	3. **Task:** Issue the text and allow pairs to discuss which articles to insert, if any.
	4. **Post-task:** The learners read their text and justify the inclusion of any article.
	The teacher comments. The teacher may point out instances when he / she feels an article is absolutely necessary for grammatical accuracy, and when it may simply be optional. Issue the original article as a handout, for learners to compare with their own work.
Comment	This activity is an awareness-raising one and can be difficult or frustrating for some learners.
Acknowledgement	This idea is based on tasks in the Linguarama writing skills courses.

Example

Text with articles deleted

city of Frankfurt, home to European Central Bank and famous as major financial and trade fair centre, is situated on River Main in heart of one of most attractive areas of Germany.

city itself is intriguing mixture of high glass towers and old, traditional buildings, with many small pubs where visitor can taste the local "Apfelwein" (cider). Frankfurt also has great choice of theatres and magnificent Opera House which is delight to visit.

Original text

The city of Frankfurt, home to the European Central Bank and famous as a major financial and trade fair centre, is situated on the River Main in the heart of one of the most attractive areas of Germany.

The city itself is an intriguing mixture of high glass towers and old, traditional buildings, with many small pubs where the visitor can taste the local "Apfelwein" (cider). Frankfurt also has a great choice of theatres and a magnificent Opera House which is a delight to visit.

(Taken from: http://www.linguarama.com)

24 Collocation search

Language areas	SPEAKING · LISTENING · READING · WRITING · INTEGRATED SKILLS · GRAMMAR · **VOCABULARY**
Activity at a glance	**Learners search for a useful authentic text on the Web and look for relevant collocations.**
Aim	To allow learners to find some word partnerships which they need to speak about their field of business.
Level	Intermediate and above
Time	30–45 minutes
Rationale	Learner motivation will be high if they have searched for the text themselves, which they will often be able to locate better than the teacher!

Procedure

1. **Intro:** Pre-lesson work on collocation. Three suggestions are:
   - Show learners how to use the key word section of Peter Wilberg and Michael Lewis's *Business English* (1990, Heinle).
   - Do a mix and match activity with some common business collocates on cards, such as: *annual turnover, profit margin, parent company, chief executive* etc.
   - Show learners how to store vocabulary using 5 – 1 / 1 – 5 boxes. (See: *Implementing the Lexical Approach*, Michael Lewis, 1997, Heinle).

2. **Pre-computer task:** Tell the learners to search the Web for a suitable text from their field. The text should be about 1 – 2 pages long. (Check the learners can use a search engine, such as *Google.*)

3. **Task:** The learners search on the Web for a suitable text within their area, which they would like to study. They print this out.

4. **Post-task:** The learners use a highlighter to mark useful collocations. They then prepare a short presentation on their area, incorporating the selected collocates.

Acknowledgement

The central ideas in this activity owe a great debt to Michael Lewis. The Internet has in fact made these ideas even more valid due to the availability of suitable texts. From *Implementing the Lexical Approach, Putting Theory into Practice*, 1st edition by Lewis, ©1997. Reprinted with permission of Heinle, a division of Thompson Learning: www.thompsonrights.com, fax 800 730-2215.

Example

Here are some collocations / word partnerships taken from a text on waste incineration:

conciliation procedure
incineration plants
incineration of waste
co-incineration of waste
vegetable waste

This is how the learner stored some of the vocabulary:

Figure 6.1
5 – 1 box

vegetable	
radioactive	
hazardous	waste
non-hazardous	
....	

25 Presentations

Business areas	**PRESENTATIONS** REPORT WRITING ESP
Activity at a glance	**Learners use the Web as a resource in order to give a short presentation.**
Aim	To focus on accuracy and fluency through giving a presentation; to practise the sub-skill of Internet research; to work on vocabulary expansion.
Level	Intermediate – advanced
Time	60 minutes
Procedure	1. **Intro:** Start with a review of the main stages of a business presentation, i.e. introducing yourself, introducing the topics, giving the main points of the presentation, etc.
	2. **Pre-computer task:** Present the idea of researching a presentation on the Web. The topic could contain a brief history, the current situation and predictions for the future.
	3. **Task:** The learners search for useful web-sites or access the chosen web-site(s). They gather information and prepare their presentations.
	4. **Post-task:** The learners regroup and give their presentations. The teacher provides feedback, correction and reformulation.
Variation	Two groups research the same news story but from different news web-sites, e.g. the BBC vs. CNN or a big news-gathering organisation like the BBC or CNN vs. a smaller, independent on-line news service. Extended over a day or a half day, this could be a project activity.

Presentation frame

Research a subject on the Web and use the framework below to prepare a short presentation.

Introduction	Good morning ladies and gentlemen. The subject of my presentation today is: I have divided my presentation into three parts: • First of all • Next • And finally
The history of the subject (past)	
The current situation (present)	
A prediction (future)	
Finish	To sum up, I spoke about • • • Are there any questions? Thank you for your attention.

26 The new product

Business areas	**PRESENTATIONS** REPORT WRITING ESP
Activity at a glance	**Learners invent a new product. They create a presentation on the product and, if they wish, design a web-site to promote it.**
Aim	To develop fluency and accuracy; to practise core business vocabulary; to provide incidental practice of some common Internet skills.
Level	Intermediate – advanced
Time	3 hours +
Rationale	This project is based on a principle of minimum input, maximum language output. The teacher may need to have some ideas to hand if one group is uncreative.
Procedure	1. **Pre-lesson task:** Check that you have the necessary equipment required.

2. **Intro:** Brainstorm items which you can buy now, but could not buy 20 years ago. Write up the examples generated on the whiteboard. e.g. a DVD, digital camera, PDA (Personal Digital Assistant), exotic fruits, health equipment etc.

3. **Pre-computer task:** Divide the group into two teams. Each should create a new product. Tell the learners that the product can be fictitious / imaginary (e.g. a miracle drug to help with language learning!)

   Phase 1: they decide on the product

   Phase 2: they prepare a presentation of the product using one of the following means:
   - a company information sheet, using a *Word* document
   - OHT transparencies
   - a *PowerPoint* presentation
   - a web-site, created using *Word*.

   Whichever way the learners choose to present their project, they should include: a product description / picture; company details; details about price; financial details about how they will break even.

   The teacher may need to offer support at certain points. E.g. to show the group how to perform an image search using *Google,* or to encourage and help the group draw a picture of their product and scan it.

4. **Task:** Presentations of the new product web-site. This should be done to the other group. For learners presenting with *PowerPoint* or using a web-site, it would be useful to use a beamer for this stage.

Comments	**Scope:** The project can be a fairly small-scope, or develop into a longer-term project, running over several lessons.

Technology: The project can be made fairly low-tech, using OHTS and a flip-chart, or slightly higher-tech involving the use of a word-processor. Developing the site may involve using more sophisticated tools.

Level: The project can be done with pre-intermediate groups, with an input of core vocabulary and a higher level of monitoring.

Acknowledgement   This idea has been used at Linguarama for many years. There are also variations in Business English books.

27 Trends

Business areas	**PRESENTATIONS** REPORT WRITING ESP
Activity at a glance	**Learners use graphs obtained from the Web as a source of material in order to give a short presentation.**
Aim	To practise the language of trends
Level	Intermediate
Time	60 minutes
Procedure	1. **Pre-lesson task:** Print off a graph from a financial web-site. Have the URL ready to give the learners in the lesson.

2. **Intro:** Play hangman with the word: *trend*. Afterwards, check learners know the meaning of the word, and explore derivatives (e.g. *trendy*) and collocations (e.g. *current*). Draw this grid on the whiteboard and have learners come out and complete the chart with any words they know to describe movement.

Figure 6.2
Language of trends

Up	Down	Adjectives
		Adverbs

The teacher at this point should input any new vocabulary which he / she feels appropriate; deals with any problems (e.g. *rise – raise*); focuses on different tenses (e.g. *rise – rose – risen*).

3. **Pre-computer task:** Present a graph to the learners. The graph could be taken from a financial site on the Internet (see Chapter 8). Use an OHT if necessary and make a short model presentation. Tell the learners you will omit a number of words in the presentation, and you will ask them to supply the words at the end. (The teacher can make a 'nonsense sound' at each omitted word). Omit these words: *horizontal axis / vertical axis / stood at / projection (into the future)*.

After the presentation, the learners supply the missing words.

Tell the learners that you have selected an out-of-date graph. They will make a short presentation of a graph, but they need to gather the up-to-date information from the Internet. They can use their own company sites, or report on an area of interest.

4. **Task:** The learners search for a graph to present. They print it off. The graph can be photocopied onto an OHT, or the data transferred by the learners to the flip-chart or whiteboard.

5. **Post-task:** The learners give their presentations. The teacher provides feedback, correction and reformulation.

Variations

Print off the graph from the web-site and give it out to the learners.

The learners use the Web to look for an interesting graph, e.g. on their own company web-sites.

28 Report writing

Business areas	PRESENTATIONS **REPORT WRITING** ESP
Activity at a glance	**Learners use the Web as a resource in order to research data for a S.W.O.T. analysis of their company.**
Aim	To give learners practice in report writing.
Level	Intermediate – advanced
Time	60 minutes
Rationale	Learners often express a need to practise the skill of report writing.

Procedure

1. **Pre-lesson task:** Decide if the learners are to make use of word processors.

2. **Intro:** Brainstorm what makes a good report. Teacher checklist:
   - Elements of effective writing e.g. accurate grammar / appropriate vocabulary / correct spelling and punctuation / tone / style / use of connectors.
   - Elements of effective report writing e.g. use of headers / bullet points / white space.

3. **Pre-computer task:** Suggest to the learners that they use a three part approach: draft / write / review. Tell them that the Internet will be available for the gathering of information during the draft writing stage.

   Check the learners know the term S.W.O.T. Ask them to sketch out ideas for a S.W.O.T. analysis of their company.

4. **Task:** The learners research any relevant data to back up their statements, and further ideas. They could also search for data on competitors to back up 'Threats'.

5. **Post-task:** The learners use their notes to draft the analysis.

Alternative	Select a different area such as the Euro, which allows the learners to use the Web to gather data.
Comment	Works well with learners who are all from the same company.
Acknowledgement	Performing a S.W.O.T. analysis is a common business activity and is used in the *Business English Teacher's Resource Book* (Sharon Nolan / Bill Reed, 1992, Longman).

29 Investing in the future

Business areas	PRESENTATIONS REPORT WRITING **ESP**
Activity at a glance	**Learners visit a stock exchange web-site in order to explore reports on the financial situation of selected companies.**
Aims	To give practice in the 'real world' task of researching information on the Web and reporting, either verbally or in a written form; to focus on financial vocabulary and the language of trends.
Level	Intermediate – advanced
Time	45 – 60 minutes
Web-sites	London – TechMark http://www.londonstockexchange.com/techmark New York – Nasdaq http://www.nasdaq.com Frankfurt – Neue Markt http://deutsche – boerse.com/nm/index_e.htm
Rationale	This task is particularly suitable for learners working in the field of finance.
Procedure	1. **Intro:** Ask the learners about warmer questions about stock markets and investing in new technology companies. See the worksheet 'Investing in the future (1)' for example questions.
	2. **Pre-computer task:** If necessary, brainstorm vocabulary surrounding buying and selling shares. Issue the worksheet 'Investing in the future (2)'.
	3. **Task:** The learners access the web-site of a stock exchange that specialises in new technology companies, e.g. Nasdaq in New York or TechMark in London; select five companies; visit those companies' web-sites and collect information about their financial state and performance.
	4. **Post-task:** The learners regroup and present the results of their research. They explain why they chose the companies they did and why they are a good investment or not. The teacher provides feedback and correction.
Extension	The learners can return to the web-site at a later date (or dates if the course is a long one) to see whether their chosen shares have risen or fallen in value. They can modify their original statements accordingly.
Variation	The same activity could be done using any other stock market.
Comment	It is important to have tried this activity yourself so that you are familiar with the operation of the web-sites. A good lead-in is to use a computer attached to a beamer to provide a demonstration of what the learners are about to do.

Investing in the future (1)

Answer the following questions:

1. Can you name any stock exchanges that specialise in new technology companies?

2. Can you name any companies quoted on these stock exchanges?

3. Which / what type of companies would you invest in and why?

Investing in the future (2)

Follow this procedure:

- Choose a stock exchange that specialises in new technology companies and visit its web-site.
- Choose five companies you want to invest in. Enter the symbol codes in the boxes below e.g. MFST is the symbol code for Microsoft Corporation.

- Use the table below to record information about the performance of the shares of each company.
- Visit the web-site of each company and collect any other financial information you think is important. Record it in the table.
- Using this information, prepare a short presentation which explains why someone should invest in these companies at this time.

Company 1	
Current share price	
Information about recent share performance	
Most recent profit / loss figures	
Information / news in favour of investing	
Information / news against investing	

Company 2	
Current share price	
Information about recent share performance	
Most recent profit / loss figures	
Information / news in favour of investing	
Information / news against investing	

Company 3

Current share price	
Information about recent share performance	
Most recent profit / loss figures	
Information / news in favour of investing	
Information / news against investing	

Company 4

Current share price	
Information about recent share performance	
Most recent profit / loss figures	
Information / news in favour of investing	
Information / news against investing	

Company 5

Current share price	
Information about recent share performance	
Most recent profit / loss figures	
Information / news in favour of investing	
Information / news against investing	

30 Personnel search

Business areas	PRESENTATIONS REPORT WRITING **ESP**
Activity at a glance	**Learners brainstorm ideas for succeeding at interviews and then explore relevant sites in order to compare their ideas.**
Aim	To practise the language of recruitment
Level	Intermediate – advanced
Time	45 minutes
Rationale	This task is particularly suitable for learners who need to interview in English.
Procedure	

1. **Intro:** Ask the learners about the recruitment procedure in their company. Do they recruit on-line? Who interviews? What are the stages of the recruitment process? Brainstorm a list of tips for job-hunters and collate on the whiteboard.

2. **Pre-computer task:** Tell the learners they are going to submit a draft for a section on their company web-site, informing prospective job seekers on how to apply for a position.

3. **Task:** The learners gather the information. They could check sites such as: http://www.newmonday.co.uk or http://www.mindtools.com

   They gather the information necessary and draft their ideas.

4. **Post-task:** The learners report back.

Variation	Brainstorm what makes a good C.V. Do a web-search to find suitable advice. Discuss the advice found.

7 Teacher training

Chapter 7 is divided into the following areas:

1 Teacher training for the Web
2 Teacher training activities

It aims to provide teacher trainers with a number of practical ideas for training teachers in using the World Wide Web in language teaching.

1   Teacher training for the Web

1.1 Approaches to teacher training

We highly recommend a practical, hands-on approach to running teacher training. A generic structure to an experiential training session on the Web involves:

- a lead-in
- a pre-computer task
- trainees performing a task
- trainees reporting back.

1.2 Experience

It is useful for the teacher trainer to be aware of the trainees' teaching experience, and whether the session forms part of pre-service training, INSET (in-service training) or Teacher Development, and whether there is a mixture of teaching experience amongst the trainees.

The other key variable is the level of knowledge of the trainees in the areas of:

- technology (familiarity with word processing and using the keyboard, and using the Internet)
- the application of technology in language teaching.

Typically, sessions may involve:

- trainees with low Web skills / little knowledge of the Internet
- trainees with a good knowledge of the Internet, but with low knowledge of language teaching applications
- a mixture.

Trainees with a low level of knowledge appreciate input in Internet-related terminology. All training sessions should include a practical element wherever possible, and include practical teaching ideas.

1.3 Equipment

It is important to know how many computers are available before running a teacher training session. The trainer should check if there any passwords, and if there is a firewall. A beamer is invaluable for demonstrations.

2 Teacher training activities

Our ideas for teacher training sessions include introductory activities and training tasks. The introductory activities are useful in helping ascertain the level of trainees, before the main body of a training session. (See: Activities 1 - 5.) The training tasks can be used in the body of the training session. (See: Activities 6 - 10.)

Activities
1 Quadrant
2 Mind Map®
3 S.W.O.T.
4 Levels
5 Jargon busting
6 Web searching
7 Practical ideas
8 The Web as a resource
9 Web-site evaluation
10 Ideas exchange

1 Quadrant

Aim To establish trends in the extent of technological knowledge about the Internet, and about using the Internet in language teaching, for trainees and their learners.

Procedure 1. **Intro:** Draw the following diagram on the whiteboard, or display it on a flip-chart.

Figure 7.1
Quadrant

	Internet	Internet in language teaching
High		
Low		

Issue each trainee with four small stickers, as used in a meta-plan activity (see Chapter 10), two red and two green. The red stickers represent the trainees; the green stickers represent a typical group of learners.

2. **Task:** Ask the trainees to place their red and green stickers in the relevant quadrants (high or low) in column one, depending on their own level of knowledge of the Internet itself (surfing, e-mail etc) and that of a typical group of learners. Then ask the trainees to do the same in column two, depending on the level of knowledge about using the Internet in language teaching.

3. **Post-task:** Discuss the pattern which emerges. Typical outcomes include trainees stating that their learners know more than they do about the Internet.

2 Mind Map®

Aim To establish trainees' level of technological knowledge about the Internet, and about using the Internet in language teaching.

Assumption The Teacher trainer and trainees are familiar with Mind Maps®. If not, for more information see: http://www.buzancentres.com

Procedure 1. **Task:** Divide the trainees into sub-groups. Use a large piece of flip-chart paper and ask the trainees to draw a Mind Map® of the Internet. Suggested branches for the Mind Map® are: *CONTENT / USES / TERMS / PROBLEMS*.

2. **Post-task:** The trainees present their Mind Maps® to each other. Typical outcomes during the task include trainees with a higher level of knowledge explaining terms to those newer to the Internet.

3 S.W.O.T.

Aim

To share information about the Internet, and views on its use.

Procedure

1. **Task:** The trainees perform a S.W.O.T. analysis of the Internet. They give examples of each category e.g.
   - Strengths: powerful search engines
   - Weaknesses: technical unreliability
   - Opportunities: wealth of materials
   - Threats: proliferation of poor sites.

Internet S.W.O.T Analysis	
Strengths	Weaknesses
Opportunities	Threats

© Summertown Publishing 2003. This material is downloadable from http://www.summertown.co.uk

Comment

This activity can help defuse negative attitudes, as it is permissible to capture cons of using the Web, while working towards a balanced overall picture.

For more on a S.W.O.T. analysis, see: *Business English Teachers' Resource Book* (Nolan, S. & Reed, B., 1992, Longman).

4 Levels

Aim

To provide information to the teacher trainer about the level of knowledge of the trainees and to re-organise the seating in order to mix levels of knowledge for the following computer activity.

Procedure

1. Ask the trainees to give themselves the level (beginner / elementary / intermediate / advanced) of their own knowledge and experience in using the Internet.

2. The trainer asks the trainees to re-organise the way they are sitting in order to form mixed-level groups.

Variation

There are many variants on this warmer. E.g. the trainees stand up and re-arrange themselves into a line, ranging from someone with the highest level to someone with the lowest level. This starts people discussing how much they know, and what. The trainer can then simply assign a letter (A-B-C) to each trainee in turn, and then ask them to re-group according to letter. The resulting groups will be mixed.

5 Jargon busting

Aim To explain key terminology.

Procedure 1. Write on the flip-chart some common Web terms which trainees need to know and ask them to check any they are unaware of with colleagues. E.g.

> search engine ADSL hyperlink homepage IPS broadband modem cache intranet

2. Typically, more knowledgeable trainees explain the terms to the others.

Variation The terms could be issued on cards, using a handout, or presented on an OHT.

6 Web searching

Aim To allow trainees to become familiar with Web searching.

Procedure 1. **Task:** Identify likely areas which a Business English teacher will need to search for, e.g. company web-sites and specialist areas. Elicit specific sites the group of trainees wish to exploit. Issue the web-site addresses of some common search engines: http://www.lycos.com or http://www.google.com or http://www.altavista.com

Clarify what a search engine is. (Demonstrate with a beamer, if necessary.) Ask trainees to work in pairs using these addresses to look for materials etc. which they would use with their learners. The trainer may wish to monitor and show features such as search refining tools.

2. **Post-task:** The trainees report on the success of using their particular search engine. Elicit how many hits it returned. What helped them in the task? How long did it take? What can help to make searching less time-consuming?

7 Practical ideas

Aim To provide hands-on experience of using the Web, using activities which the trainees could use later with their learners.

Preparation Select a number of practical lesson activities from Chapter 6, e.g. 'Investing in the future' / 'Changes' / 'Airline'.

Procedure 1. Divide the trainees into pairs. (We recommend checking that at least one person in each pair is relatively experienced in using the computer.) Explain the rationale behind the teaching activities selected and the format: aim / level / timing / procedure / comment / alternative etc. Issue the worksheets chosen from Chapter 6.

2. Give the trainees time to work through the activities as they would use them with their learners, and to prepare a short report on: what they liked about an activity / what they would adapt.

3. The trainees report back to the group.

8 The web as a resource

Aim To gain hands-on practice in adapting materials from the Internet.

Procedure
1. Ask the trainees to consider an appropriate ESP lesson they will be teaching in the near future, with an intermediate learner or group. Check the trainees are familiar with the following features: searching / printing / copying text into a word processor.
2. Using a search engine, the trainees find and then print out a text to use with their learners. They decide how they would exploit it and make any changes as necessary. They prepare a mini-presentation on how they would exploit this material in a lesson.
3. The trainees report back to the group, giving the mini-presentation.

9 Web-site evaluation

Aim To familiarise trainees with some useful sites for recommending to their learners.

Procedure
1. Issue each pair of trainees with an appropriate web-site address such as:
   - the BBC web-site
   - a translation dictionary on the Web
   - a corporate site etc.

   For suitable addresses, see Chapter 5, Activity 14.
2. Issue the 'Company web-site evaluation sheet' (see: Chapter 5, Activity 10). Give the trainees time to complete it and be ready to summarise the benefits for their learners.
3. The trainees report back.

10 Ideas exchange

Aim To enable trainees to exchange practical ideas and experiences using the Web in Business English teaching.

Pre-session task Distribute the worksheet 'Using the Web in Business English teaching' before the session.

Procedure The trainees present their activities.

Comment This type of ideas-exchange workshop works best with trainees with some knowledge of using the Web in language learning.

Using the Web in Business English teaching

1. Write down five sites you have recommended / would recommend to your learners to help them in their study of English. Why have you suggested / would you suggest the sites?

2. Note down two or three ways in which you have used the Web with your learners. Include: level / aim / site used / activity procedure.

Part three
Reference

8 Data bank

Chapter 8 is divided into the following areas:

1 **Common ESP topic area addresses**
2 **Common company addresses**

1 Common ESP topic area addresses

The following section contains twenty business topics. For each topic we have given several Web addresses. Many of these are links to web-sites concerned with the topic and are, therefore, a rich source of information, news and authentic reading materials. Others link to something unique to the Web, such as interactive reference works and the opportunity to participate in forums.

Topic areas

Accountancy
Audit
Automotive
Banking
Energy
Engineering
Environment
Finance
Information Technology
Insurance
Law
Marketing
Medical
Personnel
Retail
Tax
Telecommunications
Tourism
Transport
ESP

Accountancy

Accountancy On-line http://www.accountancymagazine.com Articles covering accountancy, tax, audit, finance. Includes daily news, analysis and surveys of firms.

Accounting Web http://www.accountingweb.co.uk or http://www.accountingweb.com Includes a weekly questions and answers section in which users can e-mail questions to a different big name from the world of accounting.

A Report http://www.areport.com

Northcote http://www.northcote.co.uk Quick access to company reports on-line. *A Report* covers the world 1000 top companies while *Northcote* just deals with British companies. You can search by business sector, country, size and which stock market the company is listed on. An easy way to find examples of balance sheets, profit and loss accounts etc.

Finexia – Business Finance Solutions http://www.finexia.com Includes calculators for working out the cost of leasing and other company financing products. Good for practising numbers in context.

Audit

Accountancy Age – audit section http://www.accountancyage.co.uk/Practice/Audit Part of the much larger *Accountancy Age* web-site. This section gives British-slanted news about financial auditing. There is also the opportunity to sign up for regular e-mail newsletters.

Audit Net http://www.auditnet.org Offers free resources for auditors, including a list of resources, a virtual library, a newsletter and much more. There is also a description of the internal audit process.

Institute of Internal Auditors http://www.theiia.org/itaudit Articles include CAAT (Computer-Assisted Auditing Techniques) and security issues. The site includes a case study on the seven-step IT risk assessment model.

The Role of the Auditor http://news.bbc.co.uk/olmedia/1690000/audio/_1692778_rake.ram A three minute audio clip of an interview with an auditor, recorded after various business scandals. A short listening which can be used as a lead-in to a discussion about the role of the auditor as it is understood by people both inside and outside the audit profession.

Automotive

Automotive Industries On-line http://www.ai-on-line.com Contains news, features, new car reviews and directories of companies in the automotive industry. http://www.ai-on-line.com/stats/globalventure2000.asp A list of all the joint-ventures and affiliations in the automotive industry worldwide.

Society of Automotive Analysts http://www.cybersaa.org A forum for new ideas and new developments in the automotive industry. Archived newsletters, a discussion forum and transcripts of presentations by industry movers and shakers. It also has a comprehensive list of links to other related web-sites.

VW (US web-site) http://www.vw.com/build.htm Build your own car web-sites. Customise a Landrover or Volkswagen by choosing the colours and features on offer. A fun way to discuss cars as products.

Auto Choice Vehicle Advisor http://autochoiceadvisor.com Click on 'Vehicle Advisor'. This is an interactive calculator which asks you a series of questions about your requirements then makes a selection from the range of cars available on the US market. Good for practising all areas of language surrounding the car business.

Banking

American Banker On-line http://www.thebankingchannel.com A resource site for bankers offering articles on industry innovations, banking headlines, conference clips with audio clips, *PowerPoint* slides and PDF files. Free e-mail alerts.

Universal Currency Converter http://www.xe.com/ucc Allows you to convert from any currency in the world to another. Great for comparing prices or the cost of living as part of a cultural awareness discussion.

Mark Bernkopf's Central Banking Resource Centre http://adams.patriot.net/~bernkopf/index-nf.html Links to Central Banks and Finance Ministries around the world, as well as links to banking histories, training resources and much more.

Start a Bank http://www.startabank.com Web-site dedicated to those who wish to set up their own bank. Can be used as a basis for extended projects.

Energy

Energy Information Administration (EIA) – Energy Glossary http://www.eia.doe.gov/glossary/glossary_main_page.htm Definitions of terms used in EIA reports. US based but comprehensive. Glossary contains links to technical documents in PDF format. The main page has links to other energy related glossaries.

Nuclear Power Plant Demonstration http://www.ida.liu.se/~her/npp/demo.html Interactive exercise in which the user has to try and prevent a meltdown. Not at all easy. Learners can predict what will happen and report back. Good for practicing instructions.

Platts Global Energy http://www.platts.com Web-site of energy information provider. Huge resource for information about the energy industry around the world. Up-to-date news, facts, figures, charts, statistics and glossaries. Try out their interactive company history.

BP Conversion Calculator http://www.bp.com/centres/energy2002/conversioncalculator/index.asp Converts between different units of measurement used in the oil and gas industry. Good for practising numbers in a realistic context.

Engineering

PBS Building Big http://www.pbs.org/wgbh/buildingbig Part of the US Public Broadcasting Service web-site dedicated to education about civil engineering. Aimed at children but the interactive labs are great for discussing building materials and introducing all the language surrounding their use and properties. Look at the 'Forces Lab' for interactive discussion of the physics of civil engineering.

Dictionary of Technical Terms for Aerospace Use http://roland.lerc.nasa.gov/~dglover/dictionary//content.html A cross-referenced dictionary which has many links to other engineering reference web-sites as well as to scientific question and answer web-sites. A rich source of vocabulary as well as a portal to a wide range of engineering company web-sites or those offering engineering information.

How Stuff Works http://www.howstuffworks.com Ever reliable with simple, clearly illustrated texts which are a rich source of vocabulary and are often linked to further explanations. Useful models for presentations on technical topics.

Institute of Civil Engineers http://www.ice.org.uk Look in the 'School Zone' for games and activities relating to civil engineering. Includes 'Who wants to be a civil engineer?'

Environment

British Government Environment Agency http://www.environment-agency.gov.uk Science and research site looking at chemical, physical and biological systems in the environment. Use the 'Find your environment' facility to locate a map with information about what is going on in the local environment. The 'Kids K-Zone' also contains simple animations about various environmental issues. Good for vocabulary and use of elementary grammar forms.

US Environmental Protection Agency (EPA) http://www.epa.gov Key topic areas include: ecosystems, pesticides, pollutants and waste. Short descriptors of current issues can be a good source of topics.

Renewable Energy Policy Project http://www.crest.org/index.html Source of useful information and articles about hydro, wind, geo-thermal and solar power.

European Environmental Agency (EEA) http://www.eea.eu.int Look at the 'Themes'. Short texts on environmental issues with a pop-up glossary on the side of each one. The web-site also contains lots of reports in PDF format with simple colour-coding and clear charts that can be used in lessons for describing trends.

Finance

CNN Money http://money.cnn.com Excellent source of information about companies – click on a company name to see a descriptor and key financial information on its performance. Company information includes graphs. In the calculator section of the web-site is 'Moneyville', a graphic, interactive savings calculator for those who wish to save for a new car or home.

Chief Financial Officer http://www.cfo.com An on-line magazine which contains a number of channels all related to company finance, e.g. accounting and tax. News stories are up-dated regularly as are links to other Web resources such as forums and financial calculators.

Financial Services Factbook http://www.financialservicesfacts.org Source of a wide range of information, facts and figures or charts relating to the financial services industry.

Bloomberg http://www.bloomberg.com The financial glossary has over 6,000 financial terms cross-referenced with 15,000 links. Access to Bloomberg TV and radio reports of the financial world.

Information Technology	**whatis.com** http://www.whatis.com Portal to several on-line databases of technical terminology as well as news about Internet technological developments, other Internet and computer specific reference web-sites and 'Ask an expert' facility.
	PC World http://www.pcworld.com/home On-line version of the computer user magazine. Sections include their highly respected news pages, computer product reviews and information, and a 'How to' section.
	How Stuff Works – Computers http://www.howstuffworks.com/category.htm?cat=Comp Clear, simple explanations of every aspect of how personal computers work. An excellent place to look when preparing for role plays with computer help desk operators.
	World Wide Consortium http://www.w3.org The web-site of the people who discuss and agree the standards on which the Web operates. Rich source of news and technical information related to the Web.
Insurance	**Insurance Institute of London** http://www.iilondon.co.uk Has an archive of presentation slides and transcripts on insurance topics dating back to 1998.
	Insurance Information Institute http://www.iii.org Portal to a range of facts and figures or research for the international insurance industry.
	Insurance Services Network http://www.isn-inc.com Describes itself as having a mission to demystify the insurance business and to provide helpful information to consumers, industry professionals and students of insurance.
	Lloyds of London http://www.lloyds.co.uk Contains lots of information about Lloyds as an insurance market, a glossary and an interactive 3D tour of the trading floor.
Law	**law.com Dictionary** http://dictionary.law.com Quick and easy dictionary of legal terms. Use either the glossary or the search box.
	International Law Dictionary and Directory http://august1.com/pubs/dict/index.shtml Listing of definitions of words and phrases used in private and public international law with cross-references to related words. Quick reference which is easy to use and has dozens of links to other related web-sites and on-line resources.
	Law from around the World http://www.law.cornell.edu/world Resource web-site from Cornell University Law School. Comprehensive. Mainly covers constitutions, statues, court law and international law. Has links to other international law web-sites.
	Hieros Gamos http://www.hg.org Legal research centre. Thousands of links to legal information on the Web. This is the next place to go if the other web-sites here cannot help you answer a legal question.
Marketing	**Chartered Institute of Marketing** http://www.cim.co.uk The web-site of the world's largest marketing body. Subscribe to a free marketing e-journal called 'What's new in marketing?' which features short reports. There is also a knowledge centre with copious mini case-studies.

Marketing Resource Centre http://www.marketingsource.com Information about how to market a company's products or services. Has a large database of articles and press releases as well as a discussion forum.

marketingprofs.com http://www.marketingprofs.com Rich source of articles and tutorials for marketing professions. Users need to register (for free) to access most parts of the web-site.

European Advertising Standards Alliance (EASA) http://www.easa-alliance.org The EASA is a Europe-wide organisation which deals with standards and other issues in advertising across the continent. Contains news and press releases as well as 'Ad Alerts' which are complaints about advertising and responses from around Europe.

Medical

American Medical Association http://www.ama-assn.org Packed with information for health professionals and patients. Includes the 'Atlas of the Body' (a huge database of diagrams and photographs related to human physiology), medical ethics topics and new developments in treatment.

British Medical Journal http://www.bmj.com/index.dtl Includes articles on current medical issues. Searchable archives of articles on almost every medical topic imaginable. Opportunity to e-mail responses to recent articles published on the web-site.

On-line Medical Dictionary http://cancerweb.ncl.ac.uk/omd Displays contents by area. Includes biochemistry, biology, cardiology, genetics, infectious diseases, nutrition, pharmacology.

Visible Human Project http://www.nlm.nih.gov/research/visible/visible_human.html or http://www.npac.syr.edu/projects/3Dvisiblehuman Contains video cross-sections of the human body which can be viewed using most media-player plug-ins. Great for medical professionals and people working with medical scanning equipment.

Personnel

HR Guide http://www.hr-guide.com Huge range of links to on-line human resources information. Topics include selection, legal issues, job analysis, training and testing.

The Chartered Institute of Personnel and Development (CIPD) http://www.cipd.co.uk Information on key law issues relating to personnel. Has a professional news service. Includes 'People Management', the CIPD's on-line magazine which has a research centre with a searchable archive of articles.

New Monday http://www.newmonday.co.uk A comprehensive job site.

Monster http://www.monster.co.uk Advice on interview techniques, psychometric testing, writing CVs (downloadable templates available) and application letters. Also has a career clinic with a wide range of advice. Perfect for discussing recruitment and for the learner looking to find a new job.

Fortune – Careers http://www.fortune.com/careers The careers web-page of the American business magazine. Lots of information for the executive employee: a careers agony aunt, quizzes such as 'How safe is your job?' and lists of the best companies to work for. Good material for stimulating discussions.

Retail

Yahoo! Retail News http://http://biz.yahoo.com/news/retail.html Lists of articles dealing with retail issues in a wide range of companies. Good source of specific texts about the retail industry.

Retail Industry http://retailindustry.about.com Portal to dozens of hyperlinks to web-sites dealing with all aspects of the retail industry such as industry news, glossaries, research and other statistics.

Stores – Magazine of the National Retail Federation http://www.stores.org Retail related topics to stimulate discussion. The 'Top 200 Global Retailers' list is useful for comparisons and numbers practice.

HandzOn – example shopping web-sites http://www.building-web-sites.com/examplesites.cfm A wide variety of on-line shops created using *HandzOn's* software. A good place to start in any discussion or project on e-retailing.

Tax

tax.org http://www.tax.org A truly international web-site dealing with current tax issues from all around the world in an academic as well as professional manner. You can subscribe to free e-mail bulletins on a number of specialist tax issues.

Yahoo! Tax Centre http://uk.biz.yahoo.com/tax/home.html or http://taxes.yahoo.com A huge range of features including a glossary, calculators helping you to work out what you pay and make comparisons with some of the biggest earners in the UK. The American version offers similar things.

International Tax http://www.taxsites.com/international.html Features many links to country-specific sites, tax associations, EU and VAT information, tax treaties, etc.

Tax World http://www.taxworld.org Links to information about tax research, policy, jobs, courses, history and some selective translations from a variety of languages. Much of it is geared towards US taxpayers, but not all.

Telecommuni-cations

Telecommunications Industry Association http://www.tiaon-line.org List of acronyms and a glossary of telecom terms.

Compare Nokia Phones http://www.nokia.com/phones/compare/index.jsp Two interactive features. 'Find your Phone' involves answering a series of questions in order to chose the Nokia phone most appropriate to you. 'Compare Phones' allows you to make a comparison between any three phones in the Nokia range.

How Stuff Works – telephone network http://www.howstuffworks.com/telephone-image.htm Animated demonstration of how a telephone network works from *How Stuff Works*. Great for describing processes along with relevant vocabulary.

European Telecommunications Standards Institute http://www.etsi.org Lots of highly technical information about standards and who uses them.

Tourism

World Tourism Organisation http://www.world-tourism.org Articles about international tourism. Global forum for tourism policy issues and a practical source of tourism know-how.

Rough Guides http://www.roughguides.com Travel site with country information. The search facility takes you to country descriptions with a lot of information which could be imported into presentations. Sign up for an e-mailed newsletter.

Lonely Planet http://www.lonelyplanet.com Different style to the *Rough Guide* site, but similar features and benefits.

Tourism Research http://www.ratztamara.com/tourism.html Information and articles about the effects of tourism around the world. Includes the 'Ecotourism Game'.

Transport

Institute of Logistics and Transport http://www.iolt.org.uk Contains news and events, training, policy changes and implications. Lots of authentic writing on the topic.

International Maritime Organisation http://www.imo.org/HOME.html Features latest news in this area, legal and safety information, marine environmental issues and lots of other information related to the merchant shipping industry.

Department of Transport (UK) http://www.dtlr.gov.uk/transport.htm All you ever needed to know about transport policy and the rules and regulations of moving things around the UK.

Volvo Trucks http://www.volvo.com/truck Click on 'Accessorise your Truck' to choose the various extras to add to the basic model. You can rotate a 3D model of the truck and see what it looks like with the different options available. Other parts of the web-site also have information about the trucking industry in general.

ESP

Martindale's 'The Reference Desk' http://www-sci.lib.uci.edu/HSG/Ref.html A truly enormous and eclectic list of hyperlinks to web-sites covering almost every topic under the sun.

Martindale's Reference Desk: Calculators On-Line Centre
http://www-sci.lib.uci.edu/HSG/RefCalculators.html Part of the above web-site. This is an index to nearly 16,000 on-line calculators beautifully organised into clear categories.

1000 Dictionaries http://www.1000dictionaries.com Another gargantuan list of links to on-line dictionaries and glossaries. Particularly useful are the sections on business and finance; English reading and writing and - if you have time - miscellaneous.

Live Radio on the Web http://www.live-radio.net/info.shtml Comprehensive list of on-line radio stations from all over the world. Time spent systematically searching the lists will yield web-sites that you will use with and recommend to learners over and over again.

Common company addresses

The following is a list of the web-sites of some major companies and corporations. The list has been organised using categories loosely based on those used by the London Stock Exchange. The list of companies itself is based on the FT100, *The Sunday Times* best 100 companies to work for and FT500 largest companies in the world. The URLs given are for the English language version of the web-site.

Categories

Advertising and marketing
Aerospace
Auditors
Automotive
Banks: commercial
Banks: national and central
Breweries, pubs and restaurants
Chemicals
Construction
Consulting
Diversified industrials
Electricity
Electronic and electrical
Finance
Food and drink
Insurance
Information Technology
Law
Leisure and hotels
Media
Medical
Metals
Mining
Oil and gas
Personal care and household products
Pharmaceuticals and biotechnology
Printing and paper
Retail: general
Retail: food
Software
Stock markets
Telecommunications
Tobacco
Transport

Advertising and marketing

Yahoo! list of advertising and marketing companies http://uk.dir.yahoo.com/ Business_and_Economy/Business_to_Business/Marketing_and_Advertising/

Abbott Mead Vickers – UK advertising agency http://www.amvbbdo.com

DBB – international advertising and marketing group http://www.dbb.com

DoubleClick Inc – Internet advertising agency http://www.doubleclick.com

Ogilvy – international advertising group http://www.ogilvy.com

Omnicom – international group of advertising companies http://www.omnicomgroup.com

Aerospace

Yahoo! list of aerospace companies http://uk.dir.yahoo.com/Business_and_Economy/ Business_to_Business/Aerospace/

Airbus – European consortium, airliner manufacturer http://www.airbus.com/

BAE Systems – UK aircraft, aircraft components and defence manufacturer http://www.baesystems.co.uk/

Boeing – US aviation, defence and space http://www.boeing.com

Lockheed Martin – US aviation, defence and space http://www.lockheedmartin.com/

Auditors

Yahoo! list of audit companies http://uk.dir.yahoo.com/Business_and_Economy/ Business_to_Business/Financial_Services/Accounting/Firms/

Deloitte & Touche – audit and financial services http://www.dttus.com/home.asp

Ernst & Young International – audit and financial services http://www.ey.com/

KPMG International – audit and financial services http://www.kpmg.com/

PriceWaterhouseCoopers – audit and financial services http://www.pwcglobal.com/

Automotive

Yahoo! list of automotive manufacturers http://uk.dir.yahoo.com/business _and_economy/shopping_and_services/automotive/makers/vehicles/

BMW – German car manufacturer http://www.bmw.com/

Citroën – French car manufacturer http://www.citroen.com/site/htm/en

Daimler Chrysler – German/US car manufacturer http://www.daimlerchrysler.de/index_e.htm

Ford – US car manufacturer http://www.ford.com

Fiat – Italian car manufacturer http://www.fiat.com/

General Motors – US car manufacturer http://www.gm.com

Honda – Japanese car manufacturer http://www.honda.co.jp/english/

MG-Rover – UK car manufacturer http://www.mg–rover.com/

Mitsubishi – Japanese car manufacturer http://www.mitsubishi–motors.co.jp/inter/entrance.html

Nissan – Japanese car manufacturer http://www.nissan–global.com/EN/HOME/

Renault – French car manufacturer http://www.renault.com/gb/accueil.htm

Volkswagen – German car manufacturer http://www.volkswagen.de/home_e/konzern/index_.html

Banks: commercial	**Yahoo! list of commercial banks** http://uk.dir.yahoo.com/business_and_economy/ shopping_and_services/financial_services/banking/banks/
	ABN Amro – Dutch bank http://www.abnamro.com
	Barclays – UK bank http://www.barclays.co.uk/
	BBVA – Spanish bank http://www.bbva.com
	Citigroup – US bank http://www.citigroup.com
	Commerzbank – German bank http://www.commerzbank.com
	Co-operative Bank – UK bank http://www.co–operativebank.co.uk
	Crédit Agricole – French bank http://www.creditagricole.com/legroupe/uk/index.shtml
	Crédit Lyonnais – French bank http://www.creditlyonnais.com/english/home/home_en.jsp
	Crédit Suisse – Swiss bank http://www.creditsuisse.com/en/home.html
	Deutsche Bank – German bank http://group.deutsche–bank.de/ghp/index_e.htm
	HSBC – UK bank http://www1.hsbc.com
	J P Morgan Chase – US bank http://www.jpmorganchase.com
	Royal Bank of Scotland – UK bank http://www.royalbankscot.co.uk/
	Société Générale Group – French bank http://www.socgen.com/en/html/index_en.htm
	UBS – Swiss bank http://www.ubs.com
	UniCredito Italiano – Italian bank http://www.unicredito.it/en/home.php3
Banks: national and central	**Bank of Canada** http://www.bank–banque–canada.ca/english/intro–e.htm
	Deutsche Bundesbank http://www.bundesbank.de/index_e.html
	Bank of England http://www.bankofengland.co.uk/
	European Bank for Reconstruction and Development http://www.ebrd.com/english/index.htm
	Bank of France http://www.banque–france.fr/gb/home.htm
	Hong Kong Monetary Authority http://www.info.gov.hk/hkma/
	Bank of Japan http://www.boj.or.jp/en/index.htm
	Federal Reserve Bank of New York http://www.ny.frb.org/
	Swiss National Bank http://www.snb.ch/
Breweries, pubs and restaurants	**Yahoo! list of breweries, pub and restaurant companies** http://uk.dir.yahoo.com/business_and_economy/shopping_and_services/food_and_drink/drinks/ alcohol_and_spirits/beer/breweries_and_brands/by_region/countries/
	McDonalds – US fast food group http://www.mcdonalds.com
	Six Continents – international hospitality group, restaurants and bars http://www.sixcontinents.com
	PizzaExpress – UK pizzeria restaurants http://www.pizzaexpress.com

Chemicals

Yahoo! list of chemical companies http://uk.dir.yahoo.com/Business_and_Economy/ Business_to_Business/Chemicals_and_Allied_Products/

Associated Octel – US chemicals producer http://www.octel–corp.com

BASF – German chemicals producer http://www.basf.de/en/

Bayer – German chemicals producer http://www.bayer.com

BOC Group – UK chemicals producer http://www.boc.com

Dow Chemical – US chemicals producer http://www.dow.com

Du Pont – US chemicals producer http://www.dupont.com

ICI – UK chemicals producer http://www.ici.com/

Henkel – German chemicals producer http://www.henkel.com

Construction

Yahoo! list of construction companies http://uk.dir.yahoo.com/business_and_economy/ business_to_business/construction/by_region/countries/

Balfour Beatty – UK construction company www.balfourbeatty.com

Kier Group – UK construction company http://www.kier.co.uk

LaFarge – French building materials producer http://www.bluecircle.co.uk/

M J Gleeson – UK construction company http://www.mjgleeson.com

Consulting

Yahoo! list of computer consultants http://uk.dir.yahoo.com/Business_and_Economy/ Business_to_Business/Computers/Services/Consulting/

Yahoo! list of management consultants http://uk.dir.yahoo.com/Business_and_Economy/ Business_to_Business/Corporate_Services/Consulting/Management_Consulting/

Accenture – international management consultant http://www.accenture.com

AIT – UK computer consultant http://www.ait.co.uk

Arup – UK engineering consultant http://www.arup.com

Bain & Company – international management consultant http://www.bain.com

CMG – UK computer consultant http://www.cmg.co.uk

PA Consulting – UK management consultant http://www.paconsulting.com

WCI – UK management consultant http://www.wcigroup.com

Diversified industrials

Yahoo! list of diversified industrial companies http://uk.dir.yahoo.com/ Business_and_Economy/Business_to_Business/Conglomerates_and_Diversified_Operations/

EON – German diversified industrial company http://www.eon.com

Mitsubishi – Japanese diversified industrial company http://www.mitsubishi.com

RWE – German diversified industrial company http://www.rwe.com/en/

Suez – French diversified industrial company http://www.suez.com

Electricity

Yahoo! list of electricity companies http://uk.dir.yahoo.com/Business_and_Economy/ Business_to_Business/Energy/Electricity/Utilities/

International Power Plc – UK electricity producer http://www.internationalpowerplc.com/

National Grid Group – UK electricity distribution http://www.nationalgrid.com/

Powergen – UK power supplier http://www.powergen.co.uk/

Electronic and electrical

Yahoo! list of electronic and electrical manufacturers http://uk.dir.yahoo.com/ business_and_economy/business_to_business/electronics/makers/

ABB – Swiss electronic and electrical equipment manufacturer http://www.abb.com

Agilent – US electronics engineering company http://www.agilent.com

General Electric – US electronic and electrical equipment manufacturer http://www.ge.com

LG Electronics – Korean electronics manufacturer http://www.lg.co.kr/english

National Panasonic – Japanese electronics manufacturer http://www.panasonic.co.jp/global

Philips Electronics – Dutch electronic and electrical equipment manufacturer http://www.philips.com

Samsung – Korean electronic and electrical equipment manufacturer http://www.samsung.com

Siemens – German telecoms/electronic engineering company http://www.siemens.com

Sony – Japanese electronics manufacturer http://www.sony.co.jp/en

Toshiba – Japanese electronics manufacturer http://www.toshiba.co.jp/worldwide

Yamaha – Japanese electronics manufacturer http://www.global.yamaha.com

Finance

Yahoo! list of venture capital companies http://uk.dir.yahoo.com/business_and_economy/business_to_business/financial_services/finance_ and_receivables/financing/corporate_finance/venture_capital/

Yahoo! list of credit card companies http://uk.dir.yahoo.com/business_and_economy/ business_to_business/financial_services/finance_and_receivables/financing/credit_cards/

Yahoo! list of investment companies http://uk.dir.yahoo.com/Business_and_Economy/Finance_and_Investment/

Yahoo! list of fund management companies http://uk.dir.yahoo.com/Business_and_Economy/Shopping_and_Services/Financial_Services/ Investment_Services/Mutual_Funds/Fund_Families/

3i – international investment manager http://www.3i.com/

American Express – US financial services and travel provider http://www.americanexpress.com

Amvescap – US financial services provider http://www.amvescap.com

Chase de Vere – UK financial advisor http://www.chasedevere.co.uk

Fidelity Investments – Bermudan investment manager http://www.fidelty.co.uk

Friends Provident – UK financial services provider http://www.friendsprovident.com

Goldman Sachs – US investment bank http://www.gs.com

Morgan Stanley – US investment bank http://www.morganstanley.com

Scottish Equitable – Dutch financial services provider http://www.scoteq.co.uk

Skandia – Swedish financial services provider http://www.skandia.co.uk

Food and drink

Yahoo! list of food and drink manufacturers http://uk.dir.yahoo.com/Business_and_Economy/Business_to_Business/Food_and_Beverage/Manufacturer_and_Processing/Manufacturer/

Allied Domecq – UK drinks producer http://www.allieddomecq.co.uk/

Bacardi Martini – Spanish drinks producer http://www.bacardi.com

Cadbury Schweppes – UK food and drinks producer http://www.cadburyschweppes.com

Coca-Cola – US soft drinks producer http://www.cocacola.com/

Danone – French food producer and processor http://www.danone.com

Diageo – UK drink producer (products include Guinness) http://www.diageo.com/

Heineken – Dutch brewer http://www.heineken.com

Kraft Foods – US food producer http://www.kraftfoods.com

Nestlé SA – Swiss food producer and processor http://www.nestle.com

Unilever – UK food producer and processor http://www.unilever.com

Insurance

Yahoo! list of insurance companies http://uk.dir.yahoo.com/business_and_economy/shopping_and_services/financial_services/insurance/

Aegon – Dutch life assurance company http://www.aegon.com

Allianz – German insurance company http://www.allianz.com/home/

Allstate – US insurance company http://www.allstate.com

American International Group – US insurance company http://www.aig.com

AXA – French insurance company http://www.axa.com

Churchill – Swiss insurance company http://www.churchill.com

Generali – Italian insurance company http://www.generali.com/home_e.html

Gerling – German insurance company http://www.gerling.com/

ING – Dutch insurance company http://www.ing.com

Legal and General – UK life assurance provider http://www.legal-and-general.co.uk/

Marsh & McLennan – US insurance company http://www.marshmac.com

Munich RE – German re-insurance company http://www.munichre.com

Old Mutual – UK life assurance provider http://www.oldmutual.com/

Prudential – UK life assurance provider http://www.prudential.com/

Royal and SunAlliance – UK insurance company http://www.royalsunalliance.com

St Paul Insurance – US insurance company http://www.stpaul.com

Swiss Life – Swiss insurance company http://www.swisslife.com

Swiss Re – Swiss re-insurance company http://www.swissre.com

Zurich – Swiss insurance company http://www.zurich.com/

| Information Technology | **Yahoo! list of hardware companies** http://uk.dir.yahoo.com/business_and_economy/ business_to_business/computers/hardware/systems/manufacturer/ |

Yahoo! list of hardware companies http://uk.dir.yahoo.com/business_and_economy/ business_to_business/computers/hardware/systems/manufacturer/

Yahoo! list of networking companies http://uk.dir.yahoo.com/business_and_economy/ business_to_business/computers/communications_and_networking/

Applied Materials – US computer component manufacturer http://www.appliedmaterials.com

Cisco – US Internet networking http://www.cisco.com

Dell Computer – US computer hardware manufacturer and retailer http://www.dell.com

Ericsson – Swedish mobile phone manufacturer http://www.ericsson.com

Hewlett-Packard – US computer hardware manufacturer http://www.hp.com

IBM – US computer hardware/software manufacturer http://www.ibm.com

Intel – US computer hardware manufacturer http://www.intel.com

Motorola – US semiconductor manufacturer http://www.motorola.com

National Semiconductor – US microchip manufacturer http://www.national.com

Nokia – Finnish mobile phone manufacturer http://www.nokia.com

Sun Microsystems – US Internet networking http://www.sun.co.uk

Taiwan Semiconductor – Taiwanese semiconductor manufacturer http://www.tsmc.com.tw/english/default.htm

Texas Instruments – US semiconductor and hardware manufacturer http://www.ti.com

Law

Yahoo! list of law companies http://uk.dir.yahoo.com/Business_and_Economy/Business_to_Business/Law/Firms/

Clifford Chance – international law firm http://www.cliffordchance.com

CMS Cameron McKenna – UK law firm http://www.cmck.com

DLA – UK law firm http://www.dla.com

Eversheds – UK law firm http://www.eversheds.com

Osbourne Clarke – UK law firm http://www.osbourneclarke.com

Wragge & Co – UK law firm http://www.wragge.com

Leisure and hotels

Yahoo! list of hotel companies http://uk.dir.yahoo.com/business_and_economy/ shopping_and_services/travel_and_transportation/accommodation/by_region/countries/

Hilton Hotels – UK hotel group http://www.hilton.com

Hyatt – US hotel group http://www.hyatt.com

Marriott International – US hotel group http://www.marriott.com

Starwood – US hotel group owns the Sheraton group http://www.starwood.com

Six Continents Hotel – US hotel group owns Holiday Inn, Inter-Continental http://www.sixcontinentshotels.com

Whitbread Hotels – UK hotel group http://www.whitbread.com

Media	**Yahoo! list of media companies** http://uk.dir.yahoo.com/Business_and_Economy/Business_to_Business/News_and_Media/ **AOL Time Warner** – US media group http://www.aoltimewarner.com **Bloomberg** – US financial and marker news provider http://www.bloomberg.com **Disney** – US media and entertainment group http://disney.go.com/corporate/ **EMI Group** – US entertainment group http://www.emigroup.com/ **Mediaset** – Italian media and photography group http://www.gruppomediaset.it/indexgruppo.jsp?lang=EN **News Corporation** – international newspaper, publishing and TV group http://www.newscorp.com/index2.html **Pearson** – UK publishing and media group http://www.pearson.com/ **Reuters** – international news agency http://www.reuters.com **Sky TV** – UK satellite TV supplier, subsidiary of *News International* http://www.corporate–ir.net/ireye/ir_site.zhtml?ticker=bsy.uk&script=11931&item_id='home.html' **Viacom** – US television and media group http://www.viacom.com **Vivendi** – French media and photography group http://www.vivendiuniversal.com
Medical	**Yahoo! list of medical equipment manufacturers** http://uk.dir.yahoo.com/ Business_and_Economy/Business_to_Business/Health_Care/Equipment_and_Supplies/ **Amersham Plc** – UK medical equipment producer http://www.amersham.co.uk/ **Edwards Lifesciences** – US medical and surgical supplies http://www.edwards.com/ **GE Medical Systems** – US medical equipment producer http://www.ge.com/medical/
Metals	**Yahoo! list of metal companies** http://uk.dir.yahoo.com/Business_and_Economy/ Business_to_Business/Industrial_Supplies/Materials/Metals/ **Alcoa** – US steel and other metals producer http://www.alcoa.com **Corus** – UK steel producer http://www.corusgroup.com/
Mining	**Yahoo! list of mining companies** http://uk.dir.yahoo.com/Business_and_Economy/Business_to_Business/Mining/ **Anglo American** – Anglo-South African mining company http://www.angloamerican.co.uk/ **Rio Tinto** – Anglo-Australian mining company http://www.riotinto.com/
Oil and gas	**Yahoo! list of oil and gas companies** http://uk.dir.yahoo.com/business_and_economy/business_to_business/energy/petroleum/ **BG Group** – UK gas producer http://www.bg–group.com/ **BOC Group** – UK gas producer http://www.boc.com/ **BP** – UK oil and gas producer http://www.alcoa.com **ChevronTexaco** – US oil and gas producer http://www.chevrontexaco.com

ENI – Italian oil and gas producer http://www.eni.it/english/home.html

ExxonMobil – US oil and gas producer http://www.exxonmobil.com

Gazprom – Russian gas and oil producer http://www.gazprom.ru/eng/default.htm

Norsk Hydro – Norwegian oil producer http://www.hydro.com/

Petrobras – Brazilian oil company http://www2.petrobras.com.br/ingles/index.asp

Philips Petroleum – US oil company http://www.phillips66.com

Repsol YPF – Spanish oil company http://www.repsol.com

Royal Dutch/Shell – Anglo-Dutch oil and gas producer
http://www2.shell.com/home/Framework

Statoil – Norwegian oil producer http://www.statoil.com

TotalFinaElf – French oil and gas producer http://www.totalfinaelf.com/us/html/index.htm

Wintershall AG – German oil and gas producer http://www.wintershall.basf.de

Yukos – Russian oil and gas producer http://www.yukos.com

Personal care and household products

Yahoo! list of personal care products manufacturers
http://uk.dir.yahoo.com/Business_and_Economy/Shopping_and_Services/Personal_Care/

Colgate-Palmolive – US personal care products manufacturer http://www.colgate.com

Gillette – US personal care products manufacturer http://www.gilette.com

L'Oréal – French personal care and household products manufacturer
http://www.loreal.com

Proctor & Gamble – US personal care products manufacturer http://www.pg.com

Reckitt Benckiser – UK household products manufacturer http://www.reckitt.com/

Unilever – Anglo-Dutch household products manufacturer http://www.unilever.com

Pharmaceuticals and biotechnology

Yahoo! list of pharmaceutical companies http://uk.dir.yahoo.com/Business_and_Economy/Business_to_Business/Health_Care/Pharmaceuticals/

Yahoo! list of biotechnology companies http://uk.dir.yahoo.com/Business_and_Economy/Business_to_Business/Scientific/Biology/Biotechnology/

Amgen – US pharmaceutical manufacturer http://www.amgen.com

AstraZeneca – UK pharmaceutical manufacturer http://www.astrazeneca.com/

Aventis Pharma – French pharmaceutical retailer http://www.aventis.com

Baxter – US pharmaceutical manufacturer http://www.baxter.com/

Boehringer Ingelheim – German pharmaceutical manufacturer
http://www.boehringer–ingelheim.com

Ei Lilly and Company – US pharmaceuticals manufacturer http://www.lilly.co.uk

GlaxoSmithKline – UK pharmaceutical company http://www.gsk.com/

Johnson & Johnson – US pharmaceutical company http://www.johnsonandjohnson.com

Nova Nordisk – Danish pharmaceutical company http://www.novo.dk

Novartis – Swiss pharmaceutical company http://www.novartis.com

Pfizer – US pharmaceutical company http://www.pfizer.com

Pharmacia – US pharmaceutical manufacturer http://www.pharmacia.com

Roche – Swiss pharmaceutical company http://www.roche.com

Sanofi – French pharmaceutical and biotechnology company
http://www.elf.fr/us/groupe/san/index.htm

Schering – German pharmaceutical and biotechnology company
http://www.schering.com

Printing and paper

Yahoo! list of printing companies
http://uk.dir.yahoo.com/Business_and_Economy/Business_to_Business/Printing/

Yahoo! list of paper companies http://uk.dir.yahoo.com/Business_and_Economy/
Business_to_Business/Industrial_Supplies/Pulp_and_Paper/

Stora Enso – Finnish forestry and paper company http://www.storaenso.com

Retail: general

Yahoo! list of retail companies
http://uk.dir.yahoo.com/Business_and_Economy/Shopping_and_Services/Retailers/

Boots Plc – UK pharmacy and toiletries retailer http://www.boots–plc.com/

Costco Wholesale – US retailer http://www.costco.com

Ebay – US on-line auction house http://www.costco.com

Ito-Yokado – Japanese retailer http://www.itoyokado.iyg.co.jp/iy/index1_e.htm

Kingfisher – UK retail group http://www.kingfisher.co.uk/english/index.htm

Kohls – US retailer http://www.kohls.com

Marks & Spencer – UK clothes and food retailer
http://www2.marksandspencer.com/thecompany/

Metro – German retailer
http://www.metro.de/servlet/PB/menu/–1_l2_ePRJ–METRODE–TOPLEVEL/index.html

Sears Roebuck – US retailer http://www.sears.com

Wal-Mart – US retailer http://www.walmart.com

Retail: food

Carrefour – French supermarket http://www.carrefour.com

Morrisons – UK supermarket http://www.morrisons.plc.uk/

Sainsbury – UK supermarket http://www.j–sainsbury.co.uk

Tesco – UK supermarket http://www.tesco.com

Walgreens – US supermarket http://www.walgreens.com

Software

Yahoo! list of software companies
http://uk.dir.yahoo.com/business_and_economy/business_to_business/computers/software/

Automatic Data Processing – US computing solutions supplier http://www.adp.com

Computer Associates – US software producer http://www.ca.com

EDS – US computing services supplier http://www.eds.com/

Logica – UK software producer http://www.logica.com/

Microsoft – US software producer http://www.microsoft.com

Oracle – US software producer http://www.oracle.com

SAP – German software producer http://www.sap.com/uk

Stock markets

Yahoo! list of stock markets http://uk.dir.yahoo.com/business_and_economy/finance_and_investment/exchanges/stock_exchanges/

Deutsche Börse Group – German stock market in Frankfurt http://deutsche–boerse.com/INTERNET/EXCHANGE/index_e.htm

London Stock Exchange – UK stock market http://www.londonstockexchange.com/

techMARK – UK high-tech companies stock market in London http://www.londonstockexchange.com/techmark/default.asp

Nasdaq – US high-tech companies stock market in New York http://www.nasdaq.com/

New York Stock Exchange – US stock market http://www.nyse.com

Nikkei – Tokyo Stock Market http://www.tse.or.jp/english/index.shtml

Telecommuni-
cations

Yahoo! list of telecommunication companies http://uk.dir.yahoo.com/Business_and_Economy/Business_to_Business/Communications_and_Networking/Telecommunications/

AT&T – US telecom company http://www.att.com

Bellsouth – US telecom company http://www.bellsouth.com

BT Group – UK telephone provider http://www.groupbt.com

Cable & Wireless – US telephone provider http://www.cw.com

China Mobile, H.K. – Hong Kong telecom company http://www.chinamobile.com/english/english.htm

Deutsche Telekom – German telecom company http://www.deutschetelekom.com/dtag/home/portal/0,14925,E,00.html

France Télécom – French telephone network http://www.francetelecom.fr/vanglais/home/homev4.html

mmO2 – UK telecom company http://www.mmo2.com/docs/home.html

NTT – Japanese telecom company http://www.ntt.com/index–e.html

NTT DoCoMo – Japanese telecom company http://www.nttdocomo.com

Olivetti – Italian telecommunication services http://www.olivetti.com

One 2 One – German mobile phone operator http://www.one2one.co.uk

Orange – French mobile phone network http://www.orange.com

Portugal Telecom – Portuguese telephone company http://www.telecom.pt/uk

Telecom Italia – Italian telecom company http://www.telecomitalia.it/index_uk.asp

Telefónica – Spanish telephone company http://www.telefonica.com/home_eng.html

Vodaphone – UK telecom company http://www.vodaphone.com

Tobacco

Yahoo! list of tobacco companies http://uk.dir.yahoo.com/Business_and_Economy/
Business_to_Business/Agriculture/Crops_and_Soil/Specific_Crops/Tobacco/

British American Tobacco – UK cigarette manufacturer http://www.bat.com/

Gallaher Group – UK cigarette manufacturer http://www.gallaher–group.com/

Imperial Tobacco – UK cigarette manufacturer http://www.imperial–tobacco.com/

Philip Morris – US cigarette manufacturer http://www.philipmorris.com

Transport

Yahoo! list of shipping companies http://uk.dir.yahoo.com/business_and_economy/
business_to_business/transportation/maritime/

Yahoo! list of trucking companies http://uk.dir.yahoo.com/business_and_economy/
business_to_business/transportation/trucks/trucking/

Yahoo! list of railway companies http://uk.dir.yahoo.com/business_and_economy/
business_to_business/transportation/trains_and_railroads/

British Airport Authority – UK airports http://www.baa.co.uk/

Exel Supply – UK transport group http://www.exel.com/

Maersk – Danish shipping http://www.maersk.co.uk

P&O – UK shipping and cruise liners http://www.pogroup.com

Railtrack – UK railway http://www.railtrack.co.uk/

9 How to... : a collection of tips

Chapter 9 is divided into the following areas (how to . . .):

The following tips are ideas and techniques which we would like to pass on. They have found their way into our everyday use of the Internet both in and out of teaching. They are ways of making the Internet more accessible, especially to those who are relatively new to it, and they make life on-line easy to organise and keep in perspective.

1 Copy information from web-pages to other programs

It is possible to copy any part of a web-page to a word processor document or to a image-manipulation program. This process can either be done by taking the entire contents of the web-page or by selecting elements individually.

To copy the entire contents – text, pictures, buttons and links – of a web-page to a word processor program do the following:

- press Crtl+A on your keyboard. This is an almost universal keyboard shortcut for 'Select All'. Every item in the web-page will become highlighted
- click on 'Edit > Copy' from the browser's menu
- go to your word processor program and open a new, blank document
- either click on 'Edit > Paste' or the 'Paste' button on the toolbar. The text and pictures from the web-page will be pasted into the document.

You may discover that the formatting which held the content in place in the browser window does not work on the word processor page. Also, not all word processor packages perform this operation in the same way. Microsoft *Word* usually successfully copies everything. Corel *WordPerfect* is often unable to transport both text and images simultaneously. Older versions of word processor programs may not recognise Web content such as hyperlinks.

If you wish to only copy certain elements of the web-page to your word processor do the following:

For text:

- use your mouse to highlight the text in the same way you do in a word processor package. Some text on web-pages, such as decorative titles, is not editable text but a computer graphic image which contains text
- use the 'Copy' command
- Paste the results into your word processor. Once this is done you should be able to edit the text, change the font and its size and colour

For pictures:

- right-click on the picture you wish to copy
- select 'Copy' from the menu that appears
- 'Paste' the picture into your word processor or image-manipulation software.

2 Save web-pages

There are a few ways of saving a web-page or the information from a web-page to use at a later time. However, there are a couple of things which should be mentioned beforehand. The first is the issue of copyright which is discussed below. It is important to consider this if you intend to use text and/or images from a web-site in teaching materials that will carry your name or be widely disseminated. The second point is regarding the transfer of web-page content to paper, especially via a word processing package. You should remember that Web and print are different media and are treated as such by the designers who work in them. This means that something that looks fabulous when displayed in a browser window may fall to pieces completely when copied to a word processor page.

The easiest way to do this is to click on 'File > Save as . . .' in your browser. Once you have chosen where you wish to save it on your hard drive, click on 'Save'. The browser will create two things: an HTML file and a folder. These will both have the same name, either the name you chose for the file or the original title of the web-page, which is normally displayed in the title bar at the top of the browser window. The HTML file is the code for formatting the web-page while the folder contains any images, JavaScripts and plug-in files needed to produce the page. Double-clicking on the HTML file will open the web-page in your browser.

An important thing to note is what happens if you save several web-pages from the same web-site. Although the HTML file contains the links between these web-pages, the links are to the web-pages on the Web. Once the history of your original visit to these web-pages on-line expires, you will not be able to use the links of the saved web-pages to travel between them. In order to do this you would have to re-write their HTML to direct the hyperlinks to the files on your hard drive instead of on the Web. Unless these are web-pages that you intend to use very often and you are not concerned about up-dating, the effort outweighs the benefits.

Another type of file that you might want to save are Adobe *Acrobat*® files. These are often called PDF files after their file suffix, for example, *Annual_Report.pdf*. PDF files can be quite large and you may not want to download them regularly. Therefore, saving them on your hard drive for quick, easy access is a good idea. Once the PDF file has downloaded completely (you can check this by clicking to the end of the document and choosing some random page in the middle), click on 'File > Save A Copy' from menu of the Adobe *Acrobat Reader* (not the browser's) and choose where you want to save it.

3   Browse off-line

The title 'Browse off-line' is a bit of a misnomer. Using *Internet Explorer* it is possible to view web-pages that you have visited on-line after breaking the telephone connection to your computer. This will only be a matter of interest if you are accessing the Internet in a situation in which you or the organisation are paying by the minute.

It is useful to understand how this works in order to set up your computer to take best advantage of this feature. First of all, open up *Internet Explorer* and click on the 'Tools' menu and select 'Internet options'. The following dialogue box opens up.

Figure 9.1
'Internet Options'
dialogue box

In Chapter 1 we saw how to set the homepage of your browser. Now we will look at the two other settings under the 'General' tab. The first is the 'Temporary Internet files' settings. This refers to a folder which contains all the files that make up the web-pages you have visited recently. These include HTML files, images, sound files etc. These files are retained in order to speed up download times next time you visit one of the stored web-pages. Your browser compares the stored version with the latest version on the Web. It will then take the unchanged elements from the hard drive of your computer and download those that have altered since your last visit. A good example of this is a news web-site; the graphics that make up the company logo and web-site menu are always the same while the headlines, photos and stories change by the hour. An added bonus of this system is that it is possible to view these web-pages while you are off-line.

What is stored in the 'Temporary Internet files' folder is controlled by the following factors.

The size of the folder

This can be set manually. The size is finite. When it is full the oldest files are discarded to make way for new ones. The bigger the folder the more files can be stored. The default setting is relatively small. If you have a new computer with a very large hard drive, then click on the 'Settings' button and increase the folder size to several hundred megabytes. It is also possible to move the folder out of the 'Windows' folder. The main reason for doing this would be if your computer has a partitioned hard (C) drive. This means that the C drive which contains all the program files, including the 'Windows' folder may be short of space. Moving the 'Temporary Internet files' folder to your D or E drive will allow you to increase its size without affecting the operation of other software. To move the folder always use the 'Move Folder' feature. Never manually cut and paste it from the 'Windows' folder.

Figure 9.2
'Temporary Internet files folder' settings dialogue box

The 'History' setting

As we have seen above, the 'History' retains the addresses of the web-pages that you have visited during the set period. It is this information that appears when you click on the 'History' button on the tool bar. It is also used to auto-complete URLs as you type in the 'Address' box. The contents of the 'Temporary Internet files' folder is also linked to the history. If you are off-line and click on a web-page name in the 'History' list, the browser will usually display that web-page. Clearly, the greater the number of days set for the 'History', the larger the 'Temporary Internet files' folder needs to be to hold the files for that length of time. There is no set ratio. The last controlling factor is how often you visit the Web and whether you visit the same web-pages again and again or regularly surf to new places.

4 Install plug-ins

On occasion you will find yourself trying to access a web-site created using *Flash* or a Web document saved in *Acrobat Reader*® format (PDF). If you do not have the appropriate plug-in installed on your computer, then the web-site will attempt to direct you to the web-sites of the program in question. (The two mentioned above are simply the most common.) If you have a high-speed connection it is a fairly simple exercise to follow the on-screen instructions and perform the download and install the plug-in. For most plug-ins the installation process is very simple. If you do not have a fast connection and are worried about the time required to download the software, many of the most common plug-ins can be found on computer magazine CD-ROMs. The installation process is exactly the same.

Visiting web-sites before using them in a lesson also gives you a chance to find out whether a plug-in needs to be installed on the computer in question. Obviously, if you install a plug-in on one computer, it is a good idea to install it on all the computers that you and your learners use.

5 Use Adobe *Acrobat Reader*®

Adobe *Acrobat Reader*® is a free plug-in which enables you to view a PDF file, which is a common format for the distribution of electronic documents. When you click on a PDF link *Acrobat Reader*® opens inside the browser window. The main thing to be aware of is that the *Acrobat Reader*® has its own tool bar. This means that if you wish to save, print or manipulate the PDF document you need to use these controls rather than those of the browser. See Section 2 above for information about saving PDF files on your hard drive.

Figure 9.3
Adobe Acrobat Reader® tool bar inside *Internet Explorer* window

6   Use *Internet Explorer* on an Apple *Mac*

There are versions of Microsoft's *Internet Explorer* available for the Apple *Mac* computer. These can be downloaded from Microsoft's web-site. The version you should download and install depends on the version of *Mac OS* operating system you have installed. For more information go to: http://www.microsoft.com/mac/download

7   Make desktop links to audio and video links

If you want to use a Web audio or video link regularly without having to access the web-page it lives on first, it is possible to place a shortcut to that link directly on the desktop of a computer. One example is our favourite: the BBC's news web-site. Direct access can allow learners to get straight to the World Service news summary or BBC1 television news.

Open up http://www.bbc.co.uk/worldservice then right-click on the World Service news hyperlink and select 'Copy Shortcut'. Next go to the desktop, right-click again and select 'Paste Shortcut'. A shortcut icon will be placed in the desktop. It will be named after the link, in this case: 'BBC World Service radio'. Now double-click this shortcut. If you are connected to the Internet, *RealPlayer* will open and the news summary will start to play.

© BBC – http://www.bbc.co.uk/worldservice/index.shtml

Figure 9.4
Click on the right-hand
mouse button and
select 'Copy Shortcut'

Record audio from the Web

One way to do this involves hardwiring your computer. This is not as major a job as it sounds. You will need a PC with a soundcard which has the following sockets: speaker/headphones, microphone and line-out. You will need to connect speakers or headphones to the first socket to monitor the sound. You then need to use a single audio lead to connect the line-out socket to the microphone socket. Your computer is now ready. To make your recording you need a couple of pieces of software: a media player and a recorder. The easiest recorder is the one built into *Windows*. Click on 'Start', 'Programs', 'Accessories', 'Entertainment' and then open *Sound Recorder*. The media player you use depends on the source audio. (See Chapter 1 for the different options available.)

Visit the web-site which holds the audio you wish to record. Click on the link. The appropriate media player will open and start to download the first part of the audio clip. As soon as the clip starts playing, click on the record button of your recording program. At the end of the clip, stop the recording and save the file in a standard audio format such as wave (WAV). This clip can now be played through most media player programs. If you have a network and the computer on which you placed the clip has a shared drive, you can make the clip accessible across the entire network.

One idea is to record a short news summary each day. Update it each morning and save the file with the same name thus replacing the previous one. This means that any network shortcuts you have made do not have to be changed whenever the recording is changed.

Finally, do not forget that like text and images borrowed from a web-site, copyright rules apply. See Section 17 below.

9 Set search engine preferences and use advanced features

On many search engines there is the possibility to set personnel preferences. For example, *Google* allows you to decide on the number of results it displays per page. The default setting is ten, but it is possible to increase this to twenty, thirty or even a hundred. Choosing a number higher than the default can speed up the process of sorting through results. The web-site places a cookie on your hard drive with these preferences so that when you return to the web-site they can be retained.

Search engines offer a number of so called advanced features. These are usually a more user-friendly way of using Boolean search expressions. One of the most useful, however, is the language setting. This can be used to limit the search to web-sites exclusively in English (or any other major language). This can be helpful when you are searching for information on big international companies which have a web-site for each of its main markets in the native language of that country.

10 Use a search engine cache to look at dead pages

Occasionally, when you click on a result from a search engine, the error page is displayed. This can mean that the web-page is unavailable due to a fault with the server that is hosting it or that the web-page has expired or has been deleted. In the second situation this is known as a dead link. However, all is not lost. The fact that the web-page was turned up by the search means that it is still in the search engine's index. Using *Google*, it is possible to access that cached version. This version rarely contains any images but if you need information from the web-page, it is better than nothing.

11 Set favourites as your homepage

One way of making a large system of favourites organised into a logical collection of folders easy to use is to set the favourites as your browser's homepage. Click on 'Tools' in the 'Internet Options' menu. In the homepage box enter 'c:/windows/favorites' and click on 'OK'. Now click on the 'Home' button and all your favourites folders will appear in the browser window. This is what will be displayed every time you open *Internet Explorer*. It is a good system to train learners and colleagues to use pre-existing favourites rather than search for something from scratch, thus saving time.

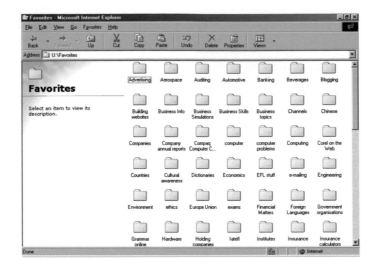

Figure 9.6
'Favourites' folders
displayed in the
Internet Explorer
window

12 **Use multiple windows to surf more quickly**

This technique allows you to have several browser windows open at once, each one downloading a different web-page. It has several advantages:

- you can keep the web-page you started on open
- you can review several web-pages quickly and decide whether they are useful or whether they can be closed
- you do not have to wait for every web-page to open, you can read one web-page while another opens in the background. This is particularly useful if you are using a modem or another type of slower Internet connection.

Once you have identified the link you want to follow, click on it using the right-hand button on your mouse. This will bring up a menu. The second item in the menu is 'Open in New Window'. Left-click this and a new browser window will open and start downloading the new web-page. You can do this as many times as you want. Move between the windows by clicking on the buttons in the task bar at the very bottom of your screen. The actual number of windows you can open depends upon the specifications of the computer. The temptation to open dozens of windows should be balanced with the amount of information you hope to find, the objectives of the lessons, the level of your learners as well as the capability of your computer to handle the process.

Figure 9.5
Using multiple
browser windows

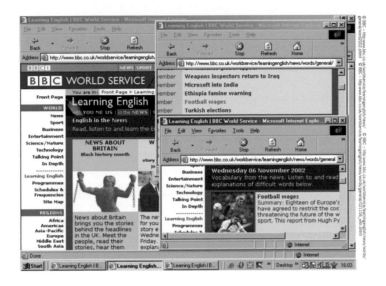

'Read' web-sites / web-pages

Throughout this book we talk about different types of web-sites. Even today the vast majority of web-sites has been created by enthusiastic amateurs. However, many of the web-sites that we use and that we suggest using have been designed, programmed and built by professionals for commercial and other types of organisation. Many of these web-sites are seen as part of the organisation's corporate image, the way it portrays itself to the outside world. Others exist to operate in the same way as commercial television or magazines do. They provide content in order to draw in an audience who will see the advertising contained within. A small, but significant, number work on a public service basis. These tend to be funded by donation or through government money. Neither of these facts detracts from the usefulness of the information contained in them.

In the world of professional web-site creation, certain unwritten standards have emerged with regard to what makes a good web-site. These concern issues such as:
• usability and navigability: can users find what they want easily and quickly?
• visual clarity: do users know what they are looking at?
• stickiness: do users stay on the web-site for very long or do they look at a couple of pages then move on?
• up-to-dateness: is the information on the web-site recent, how regularly is it changed, do users return regularly to check for new content?

In web-site design, basic principles have become established. These rules for designing a good web-site that fulfils all the criteria above can help regular Web users to become good at learning how to use web-sites and to be able to familiarise themselves quickly with new web-sites.

To give a non-Web example: British national broadsheet newspapers all follow a basic design and layout that has evolved over the years. The big story is on the front page; domestic stories precede international stories; with business, sport and arts coming towards the end of the newspaper. Editorial commentary and letters all live close together. Whether one picks up *The Guardian*, *The Times* or *The Telegraph*, finding what you want is guided by knowledge of these basic rules of layout.

The same thing has happened with many news web-sites. The main stories tend to be at the top of the homepage. The headline text is larger and there are often pictures. There is a menu, usually on the left of the browser window, listing the main sections of the web-site which usually correspond to the main sections of newspapers such as those described above. Each main story will have hyperlinks taking you to other web-pages with more details about the story; very often the background to the story in the form of previous reports and sometimes hyperlinks to other web-sites whose content has some sort of connection to the story. The web-site will also have a search facility, identifiable by a text entry box followed by a button.

Corporate homepages follow different design rules. They tend to be less cluttered. Great prominence is given to the company's logo and often images reflecting the lifestyles of the type of people the company aims its products or services at. The colours of the web-site usually reflect those of the logo and other forms of branding materials. The web-site navigation menu will have been designed and positioned wherever the designer decided was appropriate. However, it will contain a number of common items such as:

- the company's main areas of business
- the different geographical regions or countries in which the company is based
- financial information for investors
- a newsroom with company press releases
- information about the company's contributions to life outside the business world: often called 'Social Responsibility' in Europe and 'Philanthropy' on the English language versions of Japanese corporate web-sites
- information for possible new recruits to the company.

If the company is privately owned or its image is linked to that of a particular individual (see the web-site of computer manufacturer Dell) there is often biographical information about this person. These are just examples. However, the more web-sites you visit the more you will be able to recognise the main elements of this type of web-site. Another way to get an overview of these types of web-site is to look for a menu item called 'Site map'. This will give you a list of everything that is on the web-site.

What makes a good web-site?

What makes a good web-site? There are as many opinions about this as there are web-sites. However, it is a question that you should ask yourself since the web-sites you chose to use in your lessons may have an impact on the success of the lesson. A point that most designers, as well as experienced and non-experienced users agree on, is ease of navigation. If you cannot find your way around a web-site and find whether the information you require is there, then it is not much good. A rule that was established fairly early on in the life of the Web is that no page should be more than three hyperlinks away from the homepage if approached by the shortest route. This is still broken as web-sites get larger. However, if you have to click through a dozen other pages to get where you want to go, then you are unlikely to be tempted back to use that particular web-site.

In addition, remember that while you might have become used to the idiosyncratic navigation of a particular web-site over a period of time, someone using it for the first time in one of your lessons may find this departure from orthodoxy frustrating and ultimately distracting from the target of the lesson.

Discussing web-sites with learners

A first stage might be clarifying the meta-language of web-sites and web-pages. By this we mean words such as *menu, hyperlink, navigation* etc. (See Chapter 5 for learner training activities.) Our experience has shown that while learners may be very familiar with using these things, they often do not know the terms in English. It might also be useful, for non-Web initiates to demonstrate and discuss the difference between web-page content and any advertising such as banner ads and pop-ups.

A simple procedure is to ask the learners to predict what type of menu items or categories of information would be displayed on the homepage of common web-site types, such as:
- news
- corporate
- shopping
- searching tools
- government
- leisure (maybe linked to hobbies or pastimes).

Once these lists have been produced, the learners can then compare representative web-sites and report back. Here are some representative web-sites.
News web-sites: http://news.bbc.co.uk or www.cnn.com
Corporate web-sites: http://www.unilever.com or http://www.daimlerchrysler.de/index_e.htm
Searching tools: http://www.yahoo.com or http://www.wisenut.com

A good follow-up to this type of familiarisation activity is some sort of web-site assessment in which the learners use various criteria to judge how good a web-site is.

14 Deal with long download times

How do I know if there is a problem?

First of all, has your browser actually found the web-page? The easy way to check is to look in the status bar in the bottom left of the browser window. When you click on a link or type in an address and press return, the first message in the status bar reads 'Finding site' and the address of the web-site. Once the browser has contacted that web-site and starts to download the web-page, the message changes to 'Opening page' often followed by a numerical code. If this second message fails to appear then the web-page has not been found.

You can also track the progress of the little blue bar at the bottom of the browser screen which gets longer and longer as the web-page is downloaded. If this freezes, then there is also a high possibility that the web-page will not download.

Why does this happen?

There are several possible reasons for long download times, all of which, unfortunately, you can do nothing about. If there are problems with the communications lines at any point between your computer and the one you are connecting to, then things can be slowed to a snail's pace.

There are also certain times of day when things are slower. People in the United States use the Internet more than the rest of the world put together. When the east coast of America starts to wake up (5 hours behind the UK) the level of traffic on the Web can soar and web-sites that opened quickly and smoothly during the morning suddenly become lethargic. If you have become aware of a web-site which performs differently later in the day, programme a lesson which uses that web-site before eleven in the morning.

What can I do in this situation?

You cannot make a slow web-site faster. Try the 'Refresh' button on your browser (or use Ctrl+R on the keyboard) which starts the process of finding and opening the web-page again. If you have tried refreshing five times and nothing is happening, give up. Try again in an hour and you may get connected in seconds.

The best advice is to have a backup web-site or lesson plan. Obviously, if all Internet traffic is slow then going to another web-site will not help and you will have to fall back on alternative teaching methods. If your learners have any experience with using the Internet, they will appreciate the fact that sometimes you cannot open a web-page and that it is not your fault.

15   Deal with web-pages that won't open

Why does this happen?

Again, it is very unlikely to be your fault if a web-page won't open. The problem usually lies with some stage of the connection or with the web-site itself and the server on which it is stored. Browsers also give up on very slow web-pages. If the computer has been trying to download a web-page for more than about two minutes without success, it will automatically display the 'Action Cancelled' web-page.

What can I do in this situation?

If your computer fails to download a web-page:

- as before, try using the 'Refresh' button but give up after five attempts
- visit the web-sites you intend to use in a lesson before you start. If there is a problem, then you have time to find alternatives
- and you have visited the web-page recently (in the past week), then it is a connection problem. If it is a web-page you have never visited before or have not visited in several months, there is a chance that the web-page has expired and no longer exists. In this case, you need to cross it off your list or it needs to be removed from the favourites list. The link is 'dead'
- and you suspect that it has expired although you feel that it adds something vital and irreplaceable to your lesson, you can make use of *Google*'s cache feature. See Section 10 above.

16   Deal with computer viruses

Viruses are a fact of life for Internet users. Company networks have firewalls and other protection built in to protect against potential attacks, the resulting damage and cost.

The best advice to home users is to install an anti-virus program; learn how to use it and to keep its list of viruses up-to-date. These programs scan your computer and data looking for infection. However, since new viruses are released continually, it is vital to download up-dates regularly from the Internet in order for these scans to be effective and for the program to eliminate the problem.

In the event of getting a virus and not being sure how to proceed, contact the manufacturer of your chosen anti-virus software for further information.

17 Understand copyright on the Internet

Do the rules of copyright apply to Web content in the same way as they do to printed and recorded material? The clearest answer is yes they do. Many web-pages carry copyright notices and many web-sites include statements about the ownership of the content and how it may or may not be used. This applies to text, images, audio and video files as well as the HTML coding of a web-page.

A clear but robust opinion about how copyright rules apply to using information from the Web can be found at: http://whatiscopyright.org/.

See also:
http://www.benedict.com
http://www.copyright.com
http://fairuse.stanford.edu/

Much of the advice given in the above web-sites relates to the use of copyright material on the Web. However, some of the information also applies to reproduction of that material in a print-based form such as a worksheet or handout.

10 Glossary of terms

Chapter 10 is divided into the following areas:

1 **Technical terms**
2 **Pedagogical terms**

1 Technical terms

Adobe *Acrobat Reader*® Browser plug-in required for reading PDF files. The PDF file is an electronic publishing format. The text and images appear exactly as they would in a published work. This results in a more professional print-out. PDF files are a common way of distributing company brochures, reports and other published materials via the Internet. A PDF file will take longer to download than its HTML equivalent. *Acrobat Reader*® cannot be used to edit PDF documents. However, the program is free and can be downloaded from Adobe's web-site or found on CD-ROMs on the covers of computer magazines.

ADSL Stands for Asymmetric Digital Subscriber Line. A technology for sending large amounts of digital information down existing telephone lines. It can provide a permanent connection. It is described as asymmetric because it uses most of its channel to direct information to the user while the smaller, remaining proportion is used to send information out from the user. Using an ADSL, users can receive information at speeds between 512 Kbps (kilobytes per second) and 6 Mbps (megabytes per second) depending on the provider.

Alphanumeric Containing both alphabetic and numerical symbols. This is common in e-mail addresses (for example, johnsmith287@hotmail.com) and web-site URLs (for example, http://www.w3.org).

Anti-virus Software designed to detect and destroy viruses. Since new viruses are created and propagated every day, anti-virus software has to be regularly updated. This is done using an Internet connection to the software supplier. See: *Virus*.

Applet See: *Java*.

Application Some sort of computer program such as a word processor, Web browser, database etc.

Attachment See: *E-mail*.

Audio conferencing Two or more individuals in different locations having a meeting via a LAN or the Internet using microphones and headphones or speakers.

B2B Stands for Business to Business. E-commerce in which companies sell to each other using the Internet.

B2C Stands for Business to Customer / Consumer. E-commerce in which companies sell directly to end-users.

BMP Stands for bitmap. Common file format for images. This is the default file format for images created by simple graphics programs such as *Windows Paint*.

Bookmark Netscape's name for a favourite. See: *Favourite*.

Boolean Form of logic that uses terms such as *not, and* and *or*. These can be used to refine your search when using some search engines. See: *Wildcard*.

Browser Computer program used for viewing web-pages downloaded from the World Wide Web. Common examples are Microsoft *Internet Explorer, Netscape* and *Opera*. See: Chapter 1 for details.

Cache Computer memory from which high-speed retrieval of information is possible. In the case of Web browsers, the information which makes up a web-page (i.e. HTML code, images etc.) is stored in a cache on your computer's hard drive to make off-line access possible. The size of this cache can be adjusted. See: Chapter 9 for details.

Case sensitive Recognising capital letters and lower case letters as different symbols. URLs are case sensitive and so care needs to be taken in entering them. E-mail addresses are not case sensitive so it is not important whether the first letters of names are capitalised or not.

CD-R Stands for CD-Recordable. Set to be replaced by DVD-R in the not too distant future.

CD-ROM Stands for Compact Disc, Read-Only Memory. Until the recent explosion in hard drive sizes and the arrival of broadband Internet access, CD-ROMs were the best and most cost effective way of delivering multimedia content such as video, audio and large, interactive databases in the form of computer encyclopaedias. CD-ROMs can hold up to 650MB of information. The near future will see them superseded by DVD. See: *DVD*.

CD-RW Stands for CD-Rewritable. Unlike a CD-R, data on a CD-RW can be erased and the disc used again.

CGI Stands for Common Gateway Interface. This is a standard method by which information input into a web-page form is passed to an application to process, and the response to that information is passed back to the user. On-line shops use a number of CGI scripts to pass on the details of your order, name, address and credit card details to the company's computer.

Chat Synchronous text-based communication across the Internet. Chat frequently takes place in chatrooms which are part of a web-site. These chatrooms are often given over to the discussion of specific subjects. They are sometimes monitored by a human moderator who can stimulate discussion and make sure that the rules of the chatroom are followed by participants. Chat is also possible using ICQ services and software such as the Microsoft Network's *Messenger*.

Clipboard Location in the computer's memory where information can be stored temporarily. Information can be placed on the clipboard using the cut and copy commands that exist in almost all applications. This allows information to be carried from one part of a word processor document to another, as well as from one program to another. A clipboard can generally only hold one piece of information, each new piece overwriting the previous one. The clipboard retains its contents when a program is closed but not when the computer is switched off.

Compression Process by which files are reduced in size to make their transfer across a network faster. There are two types of compression. Certain file formats compress their contents. For example, an image saved in the JPEG format will be smaller than one saved in the BMP format since a JPEG is created using a compression formula. In this case, the greater the compression the greater the loss in quality of the image. Other types of compression are performed using programs such as *WinZip* which compress files of folders into a single file. The degree to which each file within a *WinZip* file is compressed depends on its format and contents.

Cookie Tiny file deposited on your computer's hard drive by a web-page. When you return to that web-page at a later date it will look for the cookie. Cookies can store information such as your browser preferences, passwords to enable you to automatically logon on to certain web-sites without entering usernames and passwords. Cookies are totally benign, they can do no harm to your computer and are often programmed to expire and disappear after a set period of time.

Copy Place information, such as text or a file, on the clipboard without removing it from its original location in a document or folder. See: *Clipboard*.

Crawler See: *Search engine*.

Cut Place information, such as text or a file, on the clipboard while, at the same time, removing it from its original location in a document or folder. See: *Clipboard*.

Cyberspace Term coined by science fiction writer William Gibson in his 1984 novel *Neuromancer* (Voyager). It refers to an imaginary Internet in which all the world's information is stored and which is accessed via a direct connection to the mind of users. The term is often used to refer to the Internet of reality.

Database Organised store of data held on a computer that can be accessed using specific programs in a variety of defined ways.

Dead link Hyperlink on a web-page or in the results of a search engine that goes nowhere or produces a browser error page. This is because the web-page in question has expired or been deleted. Some web-sites that carry large numbers of links to other web-sites invite you to report any dead links you find on their pages.

Desktop What is displayed on a computer monitor when there are no programs open. It usually displays the short-cut icons which a user clicks on to start a program or to go directly to a file or web-page.

DHTML Stands for Dynamic HTML. See: *HTML*.

Dialogue box Window which opens inside a program and presents the user with a number of manually adjustable options. For example, selecting 'Print' brings up a dialogue box in which the user can select the printer to use, which pages to print and the number of copies required.

Directory See: *Web directory*.

Download Process of transferring information in a digital, computer readable form from the Internet to your own computer.

DSL Stands for Digital Subscriber Line. See: *ADSL; VDSL*.

DVD Stands for Digital Video Disc or Digital Versatile Disc which is set to replace CD-ROM in the near future. DVD is currently more familiar as a media for delivering films and other video content which is replacing video cassettes. With the capacity of 13 CD-ROMs, DVD will soon be the main portable way of transporting computer data. At the time of writing, DVD-R equipment and media are just becoming affordable to small businesses and domestic users.

E-commerce Stands for Electronic-commerce. Business performed via the Internet. This might involve submitting orders via e-mail to a web-site able to collect and transmit data in a secure manner.

E-mail Stands for Electronic-mail. The exchange of messages from computer to computer via the Internet. Usually text, e-mails can also carry any type of computer file as attachments.

E-mail address Alphanumeric code by which the destination and sender of an e-mail is identified. An e-mail address is divided into two parts. The user name and an Internet domain name. These are separated by the @ symbol which is pronounced *at*. (For example, joe.bloggs@englishweb.com.)

Extranet Part of a web-site which is protected by passwords. This allows remote access to company information by employees and, often also, customers. See: Chapter 1 for details.

FAQ Stands for Frequently Asked Question(s). A common section on web-sites designed to help users whose questions are the same or very similar to those asked repeatedly in the past.

Favourite Tiny file containing the URL of a web-page. These are created by a browser user to enable quick access to the web-page next time the user wishes to visit it. See: Chapter 1 for details.

File Collection of information which can be read by a computer program. Information created by a computer program is also a file. Every file has a name and a suffix. The suffix identifies which program a file is associated with. For example, Microsoft *Word* creates files with the suffix .doc while Corel *WordPerfect*'s files have .wpd at the end. Some files can be read by a variety of programs. The image file format .jpg (JPEG) can be read by web browsers and every image manipulation and graphics program available.

File formats Type of computer files used on the Web, of which there are several main types. This only becomes important if you intend to transfer information in digital form to or from the Web. See: *BMP; GIF; JPG; MP3; PDF; WAV*.

Firewall Software system located at the point at which a private network such as a LAN is connected to the outside world. Its job is to protect the network from outside users accessing the information and resources that reside there. It can also be used to screen information coming into the network. This usually means e-mail attachments and downloads from the Web. What is excluded by this screening process is decided by the network administrator. As a result, a firewall on a company network may make downloading software or using streaming audio and video impossible. If you are working in-company, it is best to ask about the company's firewall policy before planning any lessons using the Internet.

Flash Player Browser plug-in that is required to experience *Flash* animations. If you have *Flash Player* then you will already have seen *Flash* at work as animated homepages, logos, menus and maybe even short cartoons and abstract, animated art. If the *Flash Player* is not installed on your computer when your browser encounters a web-page where it is required, you will be prompted to download it from the web-site of Macromedia, the company that produces *Flash*. *Flash Reader* costs nothing. This plug-in is increasingly useful as *Flash* is used to make web-sites more dynamic and interactive, for example, Coca Cola's web-site. Today it is often included as part of a standard browser installation. It can also be found on the free cover CD-ROMs of many computer magazines.

Folder Virtual container used to store computer files.

Freeware Software distributed across the Internet for which the programmer asks no payment. Microsoft and Netscape's browsers are freeware.

FTP Stands for File Transfer Protocol. Early Internet system for transferring files between computers. More or less replaced by http.

GIF Most common, compressed file format for non-photographic images used on web-sites.

Hard drive Physical store in your personal computer or in the central server of a network. The capacity is measured in MB (megabytes), for example, 30MB.

Hardware Physical components of a computer such as the monitor, hard drive, printer etc.

Homepage Frontpage of a web-site. The web-page your browser takes you to if you enter the basic URL of a web-site. When you do this the browser automatically looks for a page called index.html. This is the most common file name for a homepage. On a browser interface, homepage refers to something different. See: Chapter 1 for details.

HTML Stands for Hypertext Markup Language. This is the programming code that is used to make web-pages. It is a relatively simple programming language whose basics can be learnt very quickly. If you click on the 'View' menu of your browser and select 'Source' (*Internet Explorer*) or 'Page Source' (*Netscape*) you will be able to see the HTML code of the page you have open in the browser window. HTML is now often augmented by other programming languages. A combination of HTML and Javascript is known as DHTML (Dynamic HTML) and is used to create web-pages that have interactive elements.

http Stands for Hypertext Transfer Protocol. This is a system of rules for exchanging files of all types across the World Wide Web. Hence all web-site addresses start with http://.

Hyperlink Connection from one part of a web-page to another part of that web-page, another web-page in the same web-site or to another web-site completely. See: Chapter 1 for details.

Index See: *Search engine*.

Internet Global network of interconnected computers which communicate with each other using a common computer language. See: Chapter 1 for details.

Intranet Limited to the internal network of a particular company, part of a company or institution. It is not accessible to outside users via the Internet. See: Chapter 1 for details.

ISDN Stands for Integrated Service Digital Network. System for sending digital information over conventional phone lines. Requires special adapters at each end. Capable of speeds up to 128 Kbps. See: *DSL*; *ADSL*; *VDSL*.

ISP Stands for Internet Service Provider. Company which allows computer users to connect to the Internet. ISP offers a variety of ways of paying from free services where the user pays for the phone call to un-metered access where the user pays a monthly fee. See: Chapter 1 for details.

Java Although also a programming language, Java is not the same as JavaScript. Java is used to create applets, tiny programs which are downloaded as part of web-pages and run inside browsers. Java applets should be indifferent to the type of browser being used since the language is supposed to be truly cross-platform in nature.

JavaScript Programming language used to add movement and interactivity to web-pages. For example, when you hold your mouse over a graphic hyperlink and the button changes it is a tiny JavaScript program that is acknowledging the mouse's presence in that position and responding accordingly. How JavaScript is understood has been implemented differently by different browser manufacturers. Normally web-site designers compensate for this. Occasionally they do not and a web-page that looks stunning in one browser will not open correctly in another. This is now very rare in the professionally produced web-sites of companies and corporations.

JPG or JPEG Stands for Joint Photographers Expert Group. Most common compressed file format for photographic images used on web-sites.

LAN Stands for Local Area Network. Network of connected computers in a single locality such as one branch of a business.

Link Pre-programmed connection between two things on the Web. See: *Hyperlink*.

Lurker Someone who frequents newsgroups and discussion forums but only to read and not contribute, i.e. to lurk. Not in the least negative despite the traditional dictionary meaning of the word.

Mailing list E-mail based discussion forum. A mailing list usually covers a particular topic. Members can e-mail responses to comments or queries from other members. Each submission to the forum is e-mailed to all members on the list. Another type of mailing list is simply a list of e-mail addresses to which a regular e-mail is sent. This is usually a newsletter from a web-site. Web-sites that offer this service make signing up very simple. Removing your name from the list is also straightforward and is usually outlined at the end of every e-mail.

Media player Plug-in which allows users to hear and see audio and video content on web-sites. See: Chapter 1 for details.

Mirror web-site Copy of a web-site stored on another host computer, often in a different geographical location. This allows faster access to popular web-sites. Common wisdom says you should always access the mirror web-site closest to your own geographical location.

Modem Device for connecting a computer to the Internet via a telephone line. See: Chapter 1 for details.

Monitor Piece of hardware that contains the screen on which computer information is displayed. There are two types: Cathode Ray Tube (CRT) which use the same technology as television sets and TFT or flat-panel monitors like those found on laptop computers and, increasingly, with desktop PCs.

Mouse Hand-operated cursor control device. Comes with a range of buttons with which to click and sometimes a wheel.

MP3 Common, compressed file format for audio on web-sites.

Multimedia Combination of images, animation, sounds, music and interactivity delivered via a computer. Multimedia programs are often very large and run directly from CD-ROMs or DVDs. Multimedia web-sites are also larger than average and so take longer to download.

Navigate To find your way around a particular web-site. Web-site designers spend a lot of time in the early stages of development deciding how users will navigate the web-site. This influences the design of menus, on-screen information such as titles, buttons and links.

Navigation General term to cover how users find their way around a web-site. It usually refers to the web-site's system of menus and hyperlinks. It is generally agreed that clear, effective, well-designed and thought out navigation is often the difference between a web-site being a good or a bad one.

NetMeeting Microsoft freeware download which can be used for synchronous communication via a LAN or the Internet. Participants in meetings using *NetMeeting* can communicate with each other verbally via whiteboards and keyboard chat and visually, if web-cams are used. As well as being used in business, *NetMeeting* is finding a place in language teaching, enabling teachers and learners to be geographically distant.

Netiquette Set of rules of behaviour established by early Internet pioneers. These worked well when there were a few hundred users. Now there are many millions, Netiquette often goes by the board.

Netscape Software company which produces *Netscape* browser and e-mail software. These programs are in direct competition with the equivalent Microsoft products.

Network System of computers interconnected in some way so as to facilitate the sharing of information. The Internet is a giant network. See: *LAN*; *WAN*.

Newsgroup Collection of messages. Each message has been posted by an individual user and builds into a discussion about a particular subject. They can be accessed through special newsgroup web-sites or software or an extension of a standard e-mail program.

Off-line Not connected to the Internet. Users who pay for their Internet connection by the minute (often via a modem and conventional telephone line) are often encouraged to 'go off-line' when reading a lot of information from a web-page or in order to exploit the cache facility of *Internet Explorer*. This is to save money. See: Chapter 9.

On-line Connected to the Internet.

Operating system Piece of software such as *Windows* or *Mac OX*. The operating system is what allows the hardware of the computer to talk to you and provides the environment in which your other software can run. See: *Software*.

Paste Put information, such as text or an image, previously placed on the clipboard into a file such as a word processor document. See: *Clipboard*.

PDF See: *Adobe Acrobat Reader®*.

Pixel Picture element. The smallest part of a computer monitor's screen that can be switched on or off. The resolution of computer monitors is measured in pixels, for example, 600 x 400, 800 x 600 or 1280 x 1024.

Plug-in Program that runs inside another program when it is required. See: *Adobe Acrobat Reader®*; *Flash Player*; *Media player*.

Portal Gateway for a web-site, often that of an ISP offering a wide range of content to its users.

Program See: *Software*.

RAM Stands for Read Only Memory. The working memory of a computer. The more RAM the faster a computer can process information. Usually measured in blocks of MB (megabytes). For example, 128MB or 512MB.

Search engine Web-site which offers a Web searching service. Unlike directories, search engines use programs called spiders or crawlers to visit web-pages and analyse their contents. This information is then stored in an index. It is this index that is searched when a user enters information. Some popular search engines are: *Google*, *Altavista* and *HotBot*. See: Chapter 1 for details.

Server Central computer on a network. Usually used to store all user data such as word processor files and databases as well as e-mail and other communication services. A password is required to access information on the server via a computer on the network.

Shareware Software distributed across the Internet for which the programmer demands no payment but requests a contribution if you chose to use it. Often there is a free version with limited features upgradeable to the full version for a nominal payment. Examples include: *RealPlayer, QuickTime, WinZip*.

Software Programs or applications that are run on a computer. Browsers, word processors and *Windows* itself are all software.

Soundcard Component of a computer that converts information into sound. It has connections for speakers and headphones for the user to hear sounds and music from multimedia programs and web-sites as well as a microphone so that sound data can be inputted into a computer.

Spider See: *Search engine*.

Surf Travel the World Wide Web by jumping from web-page to web-page or web-site to web-site simply by following interesting links. This is the main exploratory way of finding out what is on the Web. It can also be a great time-wasting activity!

TCP/IP Stands for Transmission Control Protocol / Internet Protocol. Set of rules which allow computers to talk to each other over a network. The Internet is the largest network of all and TCP/IP is used to control the transactions between the computers which form the Internet.

URL Stands for Uniform Resource Locator. Address of a web-page, for example, http://www.howstuffworks.com.

VDSL Stands for Very high bit-rate DSL. Extremely high-speed asymmetric connection. Possible speeds up to 52 MBps (megabytes per second).

Video conferencing Two or more individuals in different locations having a meeting via a LAN or the Internet using webcams, microphones and headphones or speakers.

Virus Small program which, once activated, deliberately causes damage to the host computer and, in many cases, then e-mails itself to other computers. The damage can range from minor irritation to complete destruction of the computer's operating system and any information stored on the hard drive. See: *Anti-virus*.

WAN Stands for Wide Area Network. Network of computers and servers or LANs across several localities, such as the different offices or branches of a business.

WAV Stands for Wave file. Most common, un-compressed file format for audio. This is the default file format for programs such as *Windows Sound Recorder*.

Web directory Web-site which provides a Web searching service. Unlike a search engine, it relies on human operators to manually create and update a cross-referenced list or directory of web-sites. It is this directory which is searched when a request is made by a user. Most famous examples include *Yahoo, Ask Jeeves*, and *Lycos*. See: Chapter 1 for details.

Web-cam Small camera usually mounted on top of a computer monitor and used to send images across the Internet as part of synchronous communication such as video conferencing. See: *Video conferencing*.

Web-page Single HTML document on the Web. Consists of text and images and, sometimes, audio or video content. Connected to other web-pages by hyperlinks. See: Chapter 1 for details.

Web-site Collection of web-pages under a single domain name. See: Chapter 1 for details.

Wildcard Symbol used in Boolean searching. ? represents a single letter while * represents any number or no letters. For example, entering *teach** into a search engine is asking it to search for *teach, teacher, teaching, teachable*, etc. See: *Boolean*.

Windows Computer operating system created by Microsoft and installed as standard on the vast majority of PCs sold around the world. There are several versions of *Windows*. In chronological order they are: *3.1, 95, NT4, 98, 98 2nd ed., 2000* and *XP*. The basis of *Windows* is a Graphical User Interface (GUI) which represents all the information stored in a computer using a series of images and icons.

Word processing Producing documents using a standard office application which are intended to be printed on paper. These include letters, reports, teaching materials, etc.

World Wide Web Place on the Internet where web-sites live. See: Chapter 1 for details.

WYSIWYG Stands for What You See Is What You Get. Applications such as those used for Desktop publishing, word processing and Web design, in which the image displayed on the screen is an exact reproduction of the final product, i.e. document, publication or web-page.

Zip Compress a file or folder containing files to a smaller size. See: *Compression*.

Zip cartridge Portable media produced by Iomega. Zip drives and cartridges became ubiquitous during the late 1990s. The cartridges come in two formats: 100MB and 250MB.

2   **Pedagogical terms**

Auditory Those learners who remember best by hearing. See: *Kinesthetic*; *Visual*.

Asynchronous Type of communication over the Internet which is not done in real time such as sending e-mails, or posting a message to a bulletin board. See: *Synchronous*.

Blended learning Delivering a language course in a mix of ways, partially through face-to-face and partially through on-line learning. The on-line delivery could be using either asynchronous or synchronous delivery modes.

Business maze An activity where learners make decisions in order to work through a 'maze' and reach a solution to a business-related problem.

CAT Stands for Cultural Awareness Training. See: *Inter-cultural research*.

CBT Stands for Computer-Based Training. In this form of training, the trainees have access to a computer screen, as well as to the trainer who is delivering the course.

C-teaching Stands for Classroom teaching.

Collocation Word partnerships. Words which frequently combine, e.g. Managing Director, annual turnover. A common feature of the language of Business English and ESP.

Continuing to Learn At the end of a Business English course the teacher may wish to provide the learners with ideas which ensure that they continue to study the language, gain exposure to English and employ any learning strategies covered on the course. These ideas could include using the Web, for example to work on listening skills through a news site.

ESP Stands for English for Special Purposes. Teaching English which contains specialised content, such as medical English or audit-specific lexis.

F2F Stands for Face-to-Face learning, i.e. not over the Internet.

Framework material Teaching material which contains white space, for learners to fill in their ideas, within a 'frame' or border. This can allow learners time to prepare the content, and hence to increase the fluency of expression of their ideas. Examples of such material can be found in the *Business English Teachers' Resource Book* (Nolan, S and Reed, B., 1992, Longman).

Grammar clinic Web-site to which learners submit a grammar enquiry. This is answered by a language teacher / expert.

Interactivity Term applied to CD-ROMs and web / computer environments wherein the user can interact with the material in some way. E.g. there is some sort of active response, such as feedback on answers selected in a quiz.

Inter-cultural training Refers to the study of the penetration of one culture by a member of another culture. A term frequently used with cross-cultural research, which deals primarily with the similarities and differences between cultures.

Kinesthetic Those learners who remember best by doing. See: *Auditory; Visual.*

Learning to Learn A common component of many business English courses which looks at individual learning styles; learning strategies; tips on how to learn effectively e.g. how to record vocabulary.

LMS Stands for Learner Management System. See: *MLE.*

Meta-plan System often used in companies for brainstorming, which involves pinning cards containing key-words on to a cork board. Particularly popular in Austria and Germany.

Mind Map® A diagrammatic representation of a complex idea or series of ideas. A trade mark patented by Tony Buzan.

MLE Managed Learning Environment. Learners taking a course through distance learning can be supported by having access to a managed learning environment. See: *LMS.*

MOO Stands for Multi-user Object Oriented. A virtual area on the Internet, where learners from different parts of the world can meet and communicate through text.

OHT Stands for Overhead Transparency. A standard term to describe the transparent A4-sized sheet used in giving a presentation with an OHP (Overhead Projector).

POLL Stands for Partially On-Line Learning. A term like blended learning sometimes used for the fusion of computer-facilitated learning and face-to-face learning.

RP Stands for Received Pronunciation and is standard British English i.e. it has no regional variation.

Simulations Extended role play which involves the learners taking part in an activity, such as a meeting or negotiation. The learners act as themselves, rather than taking on a prescribed role.

S.W.O.T. Stands for Strengths, Weaknesses, Opportunities and Threats. A S.W.O.T. analysis is a framework which allows a company to identify these four areas. It can help an organisation focus its activities into areas where it is strong and where the greatest opportunities lie. The strengths and weaknesses can be viewed as internal factors, opportunities and threats as external.

Synchronous Type of communication which is not in real time e.g. using chat. See: *Asynchronous.*

Visual Those learners who remember best by seeing. See: *Auditory; Kinesthetic.*

WBT Stands for Web-Based Training. A term used when the Web is used as a delivery mode. Sometimes, part of the course is taken face-to-face.

Webquest Inquiry-oriented activity in which information is gathered by learners from the Web.

Bibliography

Books

Dudeney, G. (2000), *The Internet and the Language Classroom,* Cambridge University Press, Cambridge.

Eastment, D. (1999), *The Internet and ELT,* Summertown Publishing, Oxford.

Ellis, G. and Sinclair, B. (1989) *Learning to Learn English,* Cambridge University Press, Cambridge.

Ellis, M. and Johnson, C. (1994) *Teaching Business English,* Oxford University Press, Oxford.

Lewis, M. (1997) *Implementing the Lexical Approach,* Heinle, Hove.

Dudley-Evans, T. and St. John M. J. (1998) *Developments in English for Specific Purposes,* Cambridge University Press, Cambridge.

Mole, J. (1996) *Mind your Manners,* Nicholas Brealey Publishing, London.

Nolan, S. and Reed, B. (1992) *Business English Teachers' Resource Book,* Longman, Harlow.

Sharma, P. (2002) *CD-ROM: A Teacher's Handbook,* (E-book), Summertown Publishing, Oxford.

Sperling, D. (1997) *The Internet Guide for English Language Teachers,* Prentice Hall Regents, New Jersey.

Teeler, D. and Gray, P. (2000) *How to use the Internet in ELT,* Longman, Harlow.

Thornbury, S. (2001) *Uncovering Grammar,* Macmillan Heinemann, Oxford.

Windeatt, S., Hardisty, D., Eastment, D. (2000) *The Internet,* Oxford University Press, Oxford.

Articles

Hoven, D. (1999) 'A model for listening and viewing comprehension in multimedia environments' in *Language Learning & Technology,* Vol.3, No. 1.

Hughes, J. (2001) 'English for rocket scientists' in *English Teaching Professional.*

Vallance, M. (1998) 'The design and use of an Internet resource for Business English learners' in *ELT Journal.* Vol. 52/1.

Presentations

O'Driscoll, N. (1999) 'The Internet in Business English' at IATEFL / BESIG International Conference BESIG, Brighton.

Caulton, D. (2001) 'Developing Listening Skills via the Internet' at IATEFL / BESIG International Conference, Berne.

Acknowledgements

With many thanks to:

Susan Lowe, our Editor, for all her excellent suggestions, guidance and support. Louis Garnade at Summertown Publishing for his energetic and enthusiastic backing of the project.

The authors and publishers would like to thank:

Peter Lawrence and Ralph Hall of Oxford Designers and Illustrators Limited for their inspired and creative design work. Peter Mays of White Space for his excellent design and art-work on the cover. David Eastment for generously writing the Foreword to the book, and allowing us to mention *The Internet in ELT* (David Eastment, 1999, Summertown Publishing). Also, Gary Motteram and Diane Slaouti, Pete's M.Ed tutors, for their feedback and comments on assignments which have influenced the content of this book.

John Burkitt and Elizabeth Clifton of Linguarama International, for their support for the project, and permission to refer to Linguarama products and use screen grabs. Anne Laws, Inma McLeish and Annie Williams of the Group Personnel and Pedagogical Unit, Linguarama, Alton for their original ideas and inspiration.

Steven Lowe at Linguarama in Stratford-upon-Avon for his generous support. María Sharma Sacristán for her original artwork. Colleagues at Linguarama Stratford for their feedback and ideas: Clare Bruce, Simon Cook, Ann Froggatt, Evelyn Ho, Katherine Long, Claire Meese, Christiane Neiss, Graham Taylor, Sylvie Watson. Colleagues across the Linguarama Group for their initial ideas and suggestions for web-site addresses: Donna Christie, Geoff Johnson, Anna Wilby, Ian Wilson.

Many people in the Business English profession whose ideas have inspired us, among them: Gavin Dudeney, Michael Lewis, Mark Powell – readers can find more framework ideas to those in Mark Powell's book *In company* (2000, Macmillan) through the web-site: http://www.macmillan.co.uk – as well as the work of Bill Reed and Sharon Nolan. Also, for suggestions for web-site addresses, James Chamberlain.

For permission to use their trade names and screen grabs, thanks to:

© Adobe Systems Incorporated. Used with express permission. All rights reserved. Adobe, Acrobat and Reader are either registered trademarks of Adobe Systems Incorporated in the United States and other countries. Particular thanks to Heather Forte; Altavista: ALTAVISTA INTERNET OPERATIONS LIMITED ("ALTAVISTA") grants the consent to make reference to the AltaVista brand name; Google / Microsoft (in accordance with corporate guidelines) / Yahoo!; Netscape web-site © 2002 Netscape Communications Corporation. Screen shot used with permission; Real player screens are: Copyright © 1995-2002 RealNetworks, Inc. All rights reserved. RealNetworks, Real.com, RealAudio, RealVideo, RealSystem, RealPlayer, RealJukebox and RealMedia are trademarks of registered trademarks of RealNetworks, Inc. Particular thanks to Evan Pham.

For their generous permission to use their trade names, hyperlinks and screen grabs: Cambridge University Press; Collins Dictionaries; Heinle, a division of Thompson Learning; Longman; Merriam-Webster on-line; Oxford University Press. Also, thanks to Patrick Goldsmith of Oxford University Press.

Macmillan for permission to use a screen grab taken from the *Macmillan English Dictionary* © Macmillan Publishers Limited 2002. Based on the *Macmillan English Dictionary* © Bloomsbury Publishing Plc 2002 Software © TEXTware A/S, Copenhagen 2002.

The publishers would like to thank the following people:

Eric Baber of NetLearn languages, for allowing us to refer to the ELTNewsletter and the ELTOC conference; Duncan Baker for allowing us to include reference to the Grammar Clinic, to the Lydbury English Centre, and to the IATEFL BESIG Forum which he maintains; Nina O'Driscoll for her ideas which led to the creation of the *BBC world news* worksheets and inspiration for the learner activity: *Changes*; Johnathan Dykes for permission to refer to the web-based language courses provided by http://www.netlanguages.com; John Hughes, for allowing us to use his quote from *English for rocket scientists*, and Nicolas Ridley (*English Teaching Professional*); Sean Keegan, Hamish Norbrook and David Thomas at the BBC for their help in authorising the screen shots; Mind-Map® is the registered trade mark of the Buzan Organisation and used with enthusiastic permission: http://www.buzancentres.com

All web-site addresses included in this book are in the public domain. The publisher has used its best endeavours to ensure that the information is up-to-date, but cannot accept any responsibility for the suitability of the content of the sites now or in the future.

Thanks to the many people and organisations who sanctioned the mention of their web-sites in this book.

Marilyn Ault, Ph.D. Director, ALTec (Advanced Learning Technologies) for permission to include the URL http://www.4teachers.org Copyright 1995–2004 ALTec, the University of Kansas. Funded by the US Department of Education Regional Technology in Education Consortium 1995–2002 awards #R302A50008 for SCR*TEC and #R302A000015 for HPR*TEC to ALTec (Advanced Learning Technologies) at the University of Kansas Center for Research on Learning. This resource does not necessarily reflect the policies of the US Department of Education; Kenneth Beare; BESIG (Business English Special Interest Group); Nick Brieger of York Associates; Pearson Brown for the link to http://www.better-english.com; Charles Darling, Professor of English, Webmaster, Capital Community College for permission to refer to his site; David Driggs at CountryReports.org; Josef Essberger, English Club; Peter Franklin of the Delta Intercultural Academy; Evan Frendo of E4B English for Business, Berlin; Randy Glasbergen, cartoonist; Ted Goff, cartoonist. Over 1,400 cartoons with business, safety and technology themes searchable by topic or keyword. Available for use in presentation, textbooks, training materials, web-sites and newsletters:

http://www.tegoff.com; Cliff Grimes, Accel-team.com; Stephen Hargrave for permission to use the homepage screen grab of english.is.it; Honda (UK); Anthony Hughes of Edufind for the link to http://www.eltjobs.com ; IATEFL (International Association of Teaching English as a Foreign Language); Janet Ireland, Goal / QPC; Charles Kelly of Internet TESL Journal; Prof M.K.C. MacMahon, Professor of Phonetics and Head of Department, University of Glasgow, The International Phonetic Association site; Dr. Ian McMaster Editor-in-Chief of Business Spotlight; Steve Meyers of Team Technology; John Mole for allowing us to hyperlink to the DEAL map on the web-site http://www.johnmole.com; David Perry, English Live – IBI Multimedia; Reuters; Fink Ross; Tesco; Ruth Vilmi, Managing Director, Ruth Vilmi Online Education Ltd; Simon Williams, Business Development Manager, UCLES EFL for permission to refer to BULATS; Mark Wood, The Language Key; Liam Wynne, LCCIEB Marketing Manager.

Whilst every effort has been made to trace copyright holders, this may in some case have been unsuccessful. The publishers apologise for any infringement or failure to acknowledge the original sources and would be glad to include any necessary correction in subsequent printings.

The authors would also like to express their personal thanks and gratitude:

Barney, to John and Mary Barrett. Pete, to María and Jade, as ever.